P9-DDY-115

SKUNK WORKS

Also by Leo Janos

YEAGER
CRIME OF PASSION

SKUNK WORKS

A Personal Memoir of My Years at Lockheed

BEN R. RICH AND LEO JANOS

LITTLE, BROWN AND COMPANY

BOSTON   NEW YORK   TORONTO   LONDON

First Paperback Edition

Library of Congress Cataloging-in-Publication Data

Rich, Ben R.
    Skunk Works : a personal memoir of my years at Lockheed / Ben R. Rich and Leo
Janos. — 1st ed.
        p.   cm.
    Includes index.
    ISBN 0-316-74330-5 (hc)   0-316-74300-3 (pb)
    1. Lockheed Advanced Development Company — History. 2. Rich, Ben R. —
Career in aeronautics. 3. Aeronautics — Research — United States — History. I.
Janos, Leo. II. Title. III. Title: Skunk Works.
  TL565.R53 1994
  338.7'623746'0973 — dc20   94-8732

10 9 8 7 6 5 4 3 2 1

MV-NY

*Published simultaneously in Canada
by Little, Brown & Company (Canada) Limited*

Printed in the United States of America

To the men and women
of the Skunk Works,
past, present, and future

CONTENTS

ACKNOWLEDGMENTS

THE AUTHORS wish to acknowledge the contributions of various friends, co-workers, and former colleagues who enriched the pages of this book with their perspectives and recollections. Special thanks to Nancy Johnson and to Lockheed's CEO, Dan Tellep, for providing access to Kelly Johnson's logbooks and to former colleagues Sherm Mullin, Jack Gordon, Ray Passon, Dennis Thompson, Willis Hawkins, and Steve Shobert, who provided their expertise and advice through the manuscript process. Thanks also to Col. (ret.) Barry Hennessey, Pete Eames, Air Force historian Richard Hallion, and Don Welzenbach of the CIA.

Numerous Skunk Workers contributed their insights and memories. Among them: Dick Abrams, Ed Baldwin, Alan Brown, Buddy Brown, Norb Budzinske, Fred Carmody, Henry Combs, Jim Fagg, Bob Fisher, Tom Hunt, Bob Klinger, Alan Land, Tony LeVier, Red McDaris, Bob Murphy, Norm Nelson, Denys Overholser, Bill Park, Tom Pugh, Jim Ragsdale, Butch Sheffield, Steven Schoenbaum, and Dave Young.

We are particularly grateful for the participation of Air Force and CIA pilots, past and present: Bob Belor, Tony Bevacqua, William Burk Jr., Jim Cherbonneaux, Buz Carpenter, Ron Dyckman, Barry Horne, Joe Kinego, Marty Knutson, Joe Matthews, Miles Pound, Randy Elhorse, Jim Wadkins, Al Whitley, and Ed Yeilding. Significant too were the contributions of the current secretary of defense, William J. Perry, and former secretaries, Donald H. Rumsfeld, James R. Schlesinger, Harold Brown, and Caspar Weinberger; also former Chairman of the Joint Chiefs of Staff General David Jones, former Air Force Secretary Don Rice, the CIA's Richard Helms, Richard Bissell, John McMahon, Albert "Bud" Wheelon, and John Parangosky; Generals Leo Geary, Larry Welch, Jack Ledford, and Doug Nelson; National Security advisers Walt Rostow and Zbigniew Brzezinski, and Albert Wohlstetter, formerly of the President's Foreign Intelligence Advisory Board.

For obtaining photos and reference sources, appreciation to Lockheed's Denny Lombard and Eric Schulzinger, Bill Lachman and Bill Working of the Central Imaging Office, Jay Miller of Aerofax, and Tony Landis. For reference material, a salute to aviation writers Chris Pocock and Paul Crickmore.

For help often above and beyond the call of duty, our gratitude to Diana Law, Myra Gruenberg, Debbie Elliot, Karen Rich, Bert Reich, Ben Cate, and in particular to my son, Michael Rich, for his insights and suggestions, and to our wives, Hilda Rich and Bonnie Janos, for their patience and support.

Finally, our affectionate appreciation to our agent, Kathy Robbins, and to our editor, Fredrica Friedman, executive editor of Little, Brown.

Los Angeles, January 1994

SKUNK WORKS

1

A PROMISING START

It's August 1979 on the scorching Nevada desert, where Marines armed with ground-to-air Hawk missiles are trying to score a "kill" against my new airplane, an experimental prototype code-named Have Blue. We in the Skunk Works have built the world's first pure stealth fighter, which is designed to evade the Hawk's powerful radar tracking. The Marines hope to find Have Blue from at least fifty miles away and push all the right buttons so that the deadly Hawks will lock on. To help them, I've actually provided Have Blue's flight plan to the missile crew, which is like pointing my finger at a spot in the empty sky and saying, "Aim right here." All they've got to do is acquire the airplane on radar, and the homing system inside the Hawk missile will do the rest. Under combat conditions, that airplane would be blasted to pieces. If that defensive system locks on during this test, our experimental airplane flunks the course.

But I'm confident that our stealth technology will prove too elusive for even this Hawk missile's powerful tracking system (capable of detecting a live hawk riding on the thermals from thirty miles away). What makes this stealth airplane so revolutionary is that it will deflect radar beams like a bulletproof shield, and the missile battery will never electronically "see" it coming. On the Hawk's tracking system, our fighter's radar profile would show up as smaller than a hummingbird's. At least, that's what I'm betting. If I'm wrong, I'm in a hell of a bind.

Half the Pentagon's radar experts think we at the Skunk Works

*have achieved a stealth technology breakthrough that will revolu-
tionize military aviation as profoundly as the first jets did. The
other half thinks we are deluding ourselves and everyone else
with our radar test claims. Those cynics insist that we are trying
to pull a fast one — that we'll never be able to duplicate on a real
airplane the spectacular low visibility we achieved on a forty-foot
wooden model of Have Blue, sitting atop a pole on a radar test
range. Those results blew away most of the Air Force command
staff. So this demonstration against the Hawk missile is the best
way I know to shut up the nay-sayers definitively. This test is "In
your face, buddy," to those bad-mouthing our technology and
our integrity. My test pilot teased me that Vegas was giving three
to two odds on the Hawk over Have Blue. "But what do those
damned bookies know?" he added with a smirk, patting my back
reassuringly.*

*Because our stealth test airplane has been under the tightest
security, we've had to deceive the Marines into thinking that the
only thing secret about our airplane is a black box it's supposed to
be carrying in its nose that emits powerful beams to deflect incoming radar. Of course, that's all bogus. No such black box aboard, no
beams involved. The invisibility comes entirely from the airplane's
shape and its radar-absorbing composite materials.*

*The missile crew will monitor the test on their radar scope
inside their windowless command van, but a young sergeant
standing beside me will be able to verify that, despite the blank
screen, an airplane indeed flew overhead. God knows what he will
think seeing our airplane in the sky, a weird diamond-shaped
UFO, looking as if it escaped from a trailer for a new George Lucas
Star Wars epic.*

*I check my watch. Eight in the morning. The temperature
already in the nineties, heading toward a predicted high of one-
twenty F. Have Blue should be well inside the missile's radar
track, heading for us. And in a few moments I spot a distant
speck growing ever larger in the milky blue sky. I watch Have*

Blue through my binoculars as it flies at eight thousand feet. The T-38 chase plane, which usually flies on its wing in case Have Blue develops problems and needs talking down to a safe landing, is purposely following miles behind for this test. The radar dish atop the van hasn't moved, as if the power has been turned off. The cluster of missiles, which normally would be swiveling in the launcher, locked on by radar to the approaching target, are instead pointing aimlessly (and blindly) toward distant mountains. The young sergeant stares in disbelief at the sightless missiles, then gapes as the diamond-shaped aircraft zips by directly above us. "God almighty," he exclaims, "whatever that thing was, sir, it sure is carrying one hell of a powerful black box. You jammed us dead."

"Looks that way." I say and grin.

I head to the command van, and a cold blast of the air-conditioning greets me as I step inside. The Marine crew is still seated around their electronic gear with stolid determination. Their scope screen is empty. They're waiting. As far as they know, nothing has yet flown into their radar net. Suddenly a blip appears. It's moving quickly west to east in the exact coordinates of Have Blue.

"Bogie acquired, sir," the radar operator tells the young captain in charge.

For a moment I'm startled, watching a moving blip that should not be. And it is big, unmistakable.

"Looks like a T-38, sir," says the operator.

I exhale. The T-38 chase plane is being acquired by their radar detection. The radar operator has no idea that two airplanes should be on his scope — not one — and that he never did pick up Have Blue as it flew overhead.

"Sorry, sir," the young captain says to me with a smug sneer. "Looks like your gizmo isn't working too good." Had this been a combat situation, the stealth fighter could have used high-precision, laser-guided bombs against the van and that smug

captain would never have known what hit him. Might have taught him a lesson in good grammar too.

The van door opens and the young sergeant steps into the dark coolness, still looking as if he had hallucinated in the desert heat — seeing with his own eyes a strange diamond apparition that his missiles failed to lock onto.

"Captain," he began, "you won't believe this . . ."

THREE AND A HALF years earlier, on January 17, 1975, I drove to work in downtown Burbank, California, as I had for the past twenty-five years, only now I parked for the first time in the boss's slot directly in front of an unmarked two-story windowless building that resembled a concrete blockhouse, in plain view of the main runway at Burbank's busy Municipal Airport. This was Lockheed's "Skunk Works," which, throughout the long, tense years of the cold war, was one of the most secret facilities in North America and high on the targeting list of the Soviet Union in the event of nuclear war. Russian satellites regularly overflew our parking lot in the midst of Lockheed's sprawling five-square-mile production complex, probably counting our cars and analyzing how busy we were. Russian trawlers, just outside territorial limits off the southern California coastline, trained powerful eavesdropping dishes in our direction to monitor our phone calls. We believed the KGB knew our key phone numbers, and computerized recording devices aboard those trawlers probably switched on when those phones rang. U.S. intelligence intercepted references to "the Skunk Works" regularly from Soviet satellite communications simply because there was no Russian translation for our colorful nickname. Our formal name was Lockheed's Advanced Development Projects.

Even our rivals would acknowledge that whoever ran the Skunk Works had the most prestigious job in aerospace. Beginning with this mild day in January, that guy was me. I was fifty years old and in the pink.

Most Skunk Workers were handpicked by our just retired leader, Kelly Johnson, one of the reigning barons of American aviation, who first joined Lockheed in 1933 as a twenty-three-year-old fledgling engineer to help design and build the Electra twin-engine transport that helped put the young company and commercial aviation on the map. By the time he retired forty-two years later, Kelly Johnson was recognized as the preeminent aerodynamicist of his time, who had created the fastest and highest-flying military airplanes in history. Inside the Skunk Works, we were a small, intensely cohesive group consisting of about fifty veteran engineers and designers and a hundred or so expert machinists and shop workers. Our forte was building a small number of very technologically advanced airplanes for highly secret missions. What came off our drawing boards provided key strategic and technological advantages for the United States, since our enemies had no way to stop our overflights. Principal customers were the Central Intelligence Agency and the U.S. Air Force; for years we functioned as the CIA's unofficial "toy-makers," building for it fabulously successful spy planes, while developing an intimate working partnership with the agency that was unique between government and private industry. Our relations with the Air Force blue-suiters were love-hate — depending on whose heads Kelly was knocking together at any given time to keep the Skunk Works as free as possible from bureaucratic interlopers or the imperious wills of overbearing generals. To his credit Kelly never wavered in his battle for our independence from outside interference, and although more than one Air Force chief of staff over the years had to act as peacemaker between Kelly and some generals on the Air Staff, the proof of our success was that the airplanes we built operated under tight secrecy for eight to ten years before the government even acknowledged their existence. Time and again, our marching

orders from Washington were to produce airplanes or
weapons systems that were so advanced that the Soviet bloc
would be impotent to stop their missions. Which was why
most of the airplanes we built remained shrouded in the deepest operational secrecy. If the other side didn't know these aircraft existed until we introduced them in action, they would
be that much farther behind in building defenses to bring
them down. So inside the Skunk Works we operated on a
tight-lipped need-to-know basis. I figured that an analyst for
Soviet intelligence in Moscow probably knew more about my
Skunk Works projects than my own wife and children.

Even though we were the preeminent research and development operation in the free world, few Americans heard of the
Skunk Works, although their eyes would light with recognition at some of our inventions: the P-80, America's first jet
fighter; the F-104 Starfighter, our first supersonic jet attack
plane; the U-2 spy plane; the incredible SR-71 Blackbird, the
world's first three-times-the-speed-of-sound surveillance airplane; and the F-117A stealth tactical fighter that many Americans saw on CNN scoring precision bomb strikes over
Baghdad during Operation Desert Storm.

These airplanes, and other Skunk Works projects that were
unpublicized, shared a common thread: each was initiated
at the highest levels of the government out of an imperative
need to tip the cold war balance of power in our direction. For instance, the F-104, nicknamed "The Missile With
the Man In It," was an incredibly maneuverable high-
performance Mach 2 interceptor built to win the skies over
Korea in dogfights against the latest high-performance Soviet
MiGs that had been giving our combat pilots fits. The U-2 spy
plane overflew the Soviet Union for four tense years until luck
ran out and Francis Gary Powers was shot down in 1960. The
U-2 was built on direct orders from President Eisenhower,
who was desperate to breach the Iron Curtain and discover
the Russians' potential for launching a surprise, Pearl Har-

bor–style nuclear attack, which the Joint Chiefs warned could be imminent.

And it is only now, when the cold war is history, that many of our accomplishments can finally be revealed, and I can stop playing mute, much like the star-crossed rabbi who hit a hole in one on the Sabbath.

I had been Kelly Johnson's vice president for advanced projects and his personal choice to succeed him when he was forced to step down at mandatory retirement age of sixty-five. Kelly started the Skunk Works during World War II, had been Lockheed's chief engineer since 1952, and was the only airplane builder ever to win two Collier Trophies, which was the aerospace equivalent of the Hollywood Oscar, and the presidential Medal of Freedom. He had designed more than forty airplanes over his long life, many of them almost as famous in aviation as he was, and he damned well only built airplanes he believed in. He was the toughest boss west of the Mississippi, or east of it too, suffered fools for less than seven seconds, and accumulated as many detractors as admirers at the Pentagon and among Air Force commanders. But even those who would never forgive Johnson for his bullying stubbornness and hair-trigger temper were forced to salute his matchless integrity. On several occasions, Kelly actually gave back money to the government, either because we had brought in a project under budget or because he saw that what we were struggling to design or build was just not going to work.

Kelly's motto was "Be quick, be quiet, be on time." For many of us, he was the only boss we had ever known, and my first day seated behind his huge desk in the big three-hundred-square-foot corner office where Kelly had commanded every aspect of our daily operations, I felt like a three-and-half-foot-tall impostor, even though my kingdom was a windowless two-story headquarters building housing three hundred engineers, supervisors, and administrators, who operated behind thick, eavesdrop-proof walls under guard and severe security restrictions in an atmosphere about as cheery as a bomb shelter. The unmarked building was adjacent to a pair of enormous production hangars, with a combined 300,000 square feet of production and assembly space. During World War II, those hangars were used to build P-38 fighters, and later on, the fleet of Lockheed Constellations that dominated postwar commercial aviation. My challenge was to keep those six football fields' worth of floor space humming with new airplane production and development. The twin giant hangars were three stories high and dwarfed four or five nearby buildings that housed our machine shops and parts factories. Aside from a guard booth that closely screened and monitored all visitors driving into our area, there were no visible signs of the restricted Skunk Works operation. Only those with a real need to know were directed to the location of our headquarters building, which had been built for Kelly in 1962. As austere as the concrete-and-steel facility was, it seemed like a palace to those fifty of us who, back in the early 1950s, had been crammed into the small drafty offices of the original Skunk Works in Building 82, less than three hundred yards away, which was an old bomber production hangar left over from World War II and still used on some of our most sensitive projects.

I enjoyed the goodwill of my colleagues because most of us had worked together intimately under tremendous pressures for more than a quarter century. Working isolated, under rules of tight security, instilled a camaraderie probably unique in the American workplace. I was Kelly's right-hand man before succeeding him, and that carried heavy freight with most of my Skunk Works colleagues, who seemed more than willing to give me the benefit of the doubt as their new boss — and keep those second guesses to a minimum for at least the first week or so. But all of us, from department heads to the janitorial brigade, had the jitters that followed the loss of a strong father figure like Clarence "Kelly" Johnson, who had taken care of us over the years and made us among the highest-paid group in aerospace, as well as the most productive and respected. *Daddy, come back home!*

I began by loosening the leash on all my department heads. I told them what they already knew: I was not a genius like Kelly, who knew by experience and instinct how to solve the most complex technical problems. I said, "I have no intention of trying to make all the decisions around here the way that Kelly always did. From now on, you'll have to make most of the tough calls on your own. I'll be decisive in telling you what I want, then I'll step out of your way and let you do it. I'll take the crap from the big wheels, but if you screw up I want to hear it first."

I left unspoken the obvious fact that I could not be taking over at a worse time, in the sour aftermath of the Vietnam War, when defense spending was about as low as military morale, and we were down to fifteen hundred workers from a high of six thousand five years earlier. The Ford administration still had two years to run, and Defense Secretary Donald Rumsfeld was acting like a guy with battery problems on his hearing aid when it came to listening to any pitches for new airplanes. And to add anxiety to a less than promising business climate, Lockheed was then teetering on the edge of corporate and moral bankruptcy in the wake of a bribery scandal, which first surfaced the year before I took over and threatened to bring down nearly half a dozen governments around the world.

Lockheed executives admitted paying millions in bribes over more than a decade to the Dutch (Crown Prince Bernhard, husband of Queen Juliana, in particular), to key Japanese and West German politicians, to Italian officials and generals, and to other highly placed figures from Hong Kong to Saudi Arabia, in order to get them to buy our airplanes. Kelly was so sickened by these revelations that he had almost quit, even though the top Lockheed management implicated in the scandal resigned in disgrace.

Lockheed was convulsed by some of the worst troubles to simultaneously confront an American corporation. We were also nearly bankrupt from an ill-conceived attempt to reenter the commercial airliner sweepstakes in 1969 with our own Tristar L-1011 in competition against the McDonnell Douglas DC-10. They used American engines, while we teamed up with Rolls-Royce, thinking that the Anglo-American partnership gave us an advantage in the European market. We had built a dozen airliners when Rolls-Royce unexpectedly declared bankruptcy, leaving us with twelve hugely expensive, engineless "gliders" that nobody wanted. The British government bailed out Rolls-Royce in 1971, and the following year Congress *very* reluctantly came to our rescue by voting us $250 million in loan guarantees; but our losses ultimately reached a staggering $2 billion, and in late 1974, Textron Corporation almost acquired all of Lockheed at a "fire sale" price of $85 million. The Skunk Works would have been sold off with the corporation's other assets and then tossed into limbo as a tax write-off.

I had to get new business fast or face mounting pressure from the corporate bean counters to unload my higher-salaried people. Kelly was known far and wide as "Mr. Lockheed." No one upstairs had dared to cross him. But I was just plain Ben Rich. I was respected by the corporate types, but I had no political clout whatsoever. They demanded that I be a hell of a lot more "client friendly" than Kelly had been. It was an open secret in the industry that Kelly had often been his own worst enemy in his unbending and stubborn dealings with the blue-suiters. Until they had run afoul of our leader, not too many two- or three-star generals had been told to their faces that they didn't know shit from Shinola. But smoothing relations with Pentagon brass would only serve to push me away from the dock — I had a long hard row ahead to reach the promised land. If the Skunk Works hoped to survive as a viable entity, we somehow would have to refashion the glory

years last enjoyed in the 1960s when we had forty-two separate projects going and helped Lockheed become the aerospace industry leader in defense contracts.

I knew there were several powerful enemies of the Skunk
Works on Lockheed's board who would close us down in a
flash. They resented our independence and occasional arro-
gance, and suspected us of being profligate spenders hiding
our excesses behind screens of secrecy imposed by our highly
classified work. These suspicions were fueled by the fact that
Kelly usually got whatever he wanted from Lockheed's
board — whether it was costly new machinery or raises for
his top people. Nevertheless, Kelly actually was as tightfisted
as any beady-eyed New England banker and would raise hell
the moment we began dropping behind schedule or going
over budget.

Knowing that I didn't have much time to find new business,
I flew to Washington, hat in hand, with a fresh shoeshine and
a brave smile. My objective was to convince General David
Jones, the Air Force chief of staff, of the need to restart the
production line of the U-2 spy plane. It was a long-shot attempt, to say the least, because never before in history had the
blue-suiters ever reopened a production line for any airplane
in the Air Force's inventory. But this airplane was special. I
have no doubt that fifty years from now the U-2 will still be in
service to the nation. The aircraft was then more than twenty-
five years old and remained the mainstay of our airborne reconnaissance activities. It needed to be updated with a more
powerful engine and fitted with advanced avionics to become
even more effective flying its tactical missions around the
world. That meant adding a capability to perform reconnaissance coverage via optical systems that used radar camera images from half a world away.

But airplanes are like people. They tend to gain weight as
they get older. The first time the U-2 took off to overfly Russia
back in 1955, it was a svelte youngster at 17,000 pounds. Now

it had ballooned in middle age to 40 percent over the original
model and bent the scales at 40,000 pounds. I had been trying
for years to get the Pentagon to update the U-2. In the 1960s, I
had a meeting with Alain Enthoven, who was head of Secretary of Defense Bob McNamara's vaunted systems analysis
group — the so-called Whiz Kids, many brought with him
from Ford to work their competitive cold-bloodedness on the
Department of Defense. Enthoven asked, "Why should we buy
more U-2s when we haven't lost any?" I explained that it was
cheaper to buy and update the airplane now rather than wait
for crashes or losses, because in ten years costs rise by a factor
of ten. He just couldn't see the logic. So I told him the story of
the kid who proudly tells his father that he saved a quarter by
running alongside a bus rather than taking it. The father
slapped the kid on the head for not running next to a taxi and
saving a buck fifty. Alain didn't get it.

During his reign, Kelly insisted on dealing with all of the
top Pentagon brass himself, so by necessity I nibbled around
the edges for years, cultivating bright young majors and colonels on the way up who were now taking command as generals. I had gone to the Pentagon many times as Kelly's chart
holder while he briefed the brass. Once we briefed
McNamara, seated behind the big desk that had belonged to
General "Black Jack" Pershing, the World War I Army general,
on our Mach 3 Blackbird spy plane, which we wanted to convert into an interceptor. It was a great idea, but we were fighting an uphill battle. McNamara was intent on buying a costly
new bomber, the B-70, and was deaf to any other new airplane
projects. I set up the charts while Kelly made the pitch during
McNamara's lunch hour. "Mac the Knife" sat concentrating
intently on his soup and salad, while skimming a report of
some sort, and never once looked up until we were finished.
Then he wiped his lips with a napkin and bid us good day. On
the way out I teased Kelly, "Never try to pitch a guy while he's
eating and reading at the same time."

Now the situation was more propitious for eating and pitching at the same time. General Jones invited me in for lunch and was very favorably disposed to my idea for a new fleet of spiffy U-2s. I told him I'd give him a good price, but that he had to buy the entire production line of forty-five airplanes. Jones thought thirty-five would be more like it and said he'd study our proposal. "By the way," he said, "I'd want the U-2 designation changed. No spy plane connotation that would make our allies shy about letting us use their bases."

I said, "General, I believe in the well-known golden rule. If you've got the gold, you make the rules. Call it whatever you want."

The Pentagon ultimately renamed the U-2 the TR-1. *T* for tactical, *R* for reconnaissance. The press immediately called it the TR-1 spy plane.

I left the Pentagon thinking we had a deal, but the study General Jones ordered took months to wend its way through the blue-suit bureaucracy, and we didn't sign the contract for two more years. Updating our old airplanes would help to keep our corporate accountants at bay for a while. With the TR-1, I was merely buying time. To survive, the Skunk Works needed substantial new projects involving revolutionary new technology that our customer could not wait to get his hands on. Tightrope walking on the cutting edge was our stock-in-trade.

"Don't try to ape me," Kelly had advised me. "Don't try to take credit for the airplanes I built. Go build your own. And don't build an airplane you don't really believe in. Don't prostitute yourself or the reputation of the Skunk Works. Do what's right by sticking to your convictions and you'll do okay."

As it happened, I was damned lucky. Stealth technology landed in my lap — a gift from the gods assigned to take care of beleaguered executives, I guess. I take credit for immediately recognizing the value of the gift I was handed before it became apparent to everyone else, and for taking major risks in expending development costs before we had any real government interest or commitment. The result was that we produced the most significant advance in military aviation since jet engines, while rendering null and void the enormous 300-billion-ruble investment the Soviets had made in missile and radar defenses over the years. No matter how potent their missiles or powerful their radar, they could not shoot down what they could not see. The only limits on a stealth attack airplane were its own fuel capacity and range. Otherwise, the means to counter stealth were beyond current technology, demanding unreasonably costly funding and the creation of new generations of supercomputers at least twenty-five years off. I felt certain that stealth airplanes would rule the skies for the remainder of my lifetime. And I came from a family of long livers.

The stealth story actually began in July 1975, about six months after I took over the Skunk Works. I attended one of those periodic secret Pentagon briefings held to update those with a need to know on the latest Soviet technical advances in weapons and electronics. The U.S. had only two defensive ground-to-air missile systems deployed to protect bases — the Patriot and the Hawk, both only so-so in comparison to the Soviet weapons.

By contrast, the Russians deployed fifteen different missile systems to defend their cities and vital strategic interests. Those of us in the business of furnishing attack systems had to be updated on the latest defensive threat. Then we would go back to the drawing board to find new ways to defeat those defenses, while the other side was equally busy devising fresh obstacles to our plans. It was point counterpoint, played without end. Their early-warning radar systems, with 200-foot-long antennas, could pick up an intruding aircraft from hundreds of miles away. Those long-range systems couldn't tell altitude or the type of airplane invading their airspace, but passed along the intruder to systems that could.

Their SAM ground-to-air missile batteries were able to en-
gage both low-flying attack fighters and cruise missiles at the
same time. Their fighters were armed with warning radars
and air-to-air missiles capable of distinguishing between low-
flying aircraft and ground clutter with disarming effectiveness. The Soviet SAM-5, a defensive surface-to-air missile of
tremendous thrust, could reach heights of 125,000 feet and
could be tipped with small nuclear warheads. At that height,
the Soviets didn't worry about impacting the ground below
with the heat or shock wave from a very small megaton
atomic blast and estimated that upper stratospheric winds
would carry the radiation fallout over Finland or Sweden. An
atomic explosion by an air defense missile could bring down
any high-flying enemy bomber within a vicinity of probably a
hundred miles with its shock wave and explosive power. Our
Air Force crews undertaking reconnaissance intelligence-
gathering missions over territory protected by SAM-5 sites all
wore special glasses that would keep them from going blind
from atomic flash. So these weapons system advances posed a
damned serious threat.

Most troublesome, the Russians were exporting their advanced nonnuclear defensive systems to clients and customers
around the world, making our airplanes and crews increasingly vulnerable. The Syrians now had nonnuclear SAM-5s.
And during our Pentagon briefing we were subjected to a chilling analysis of the 1973 Yom Kippur War involving Israel,
Syria, and Egypt. What we heard was extremely upsetting.
Although the Israelis flew our latest and most advanced jet
attack aircraft and their combat pilots were equal to our own,
they suffered tremendous losses against an estimated arsenal
of 30,000 Soviet-supplied missiles to the Arab forces. The Israelis lost 109 airplanes in 18 days, mostly to radar-guided
ground-to-air missiles and antiaircraft batteries, manned by
undertrained and often undisciplined Egyptian and Syrian
personnel. What really rattled our Air Force planners was that

the evasive maneuvering by Israeli pilots to avoid missiles — the same tactics used by our own pilots — proved to be a disaster. All the turning and twisting calculated to slow down an incoming missile made the Israeli aircraft vulnerable to conventional ground fire. If the Israeli loss ratio were extrapolated into a war between the U.S. and the highly trained Soviet Union and Warsaw Pact in Eastern Europe, a war fought using similar airplanes, pilot training, and ground defenses, our air force could expect to be decimated in only seventeen days.

I was not too surprised. The Skunk Works had firsthand experience with the latest Soviet equipment because the CIA had scored spectacular covert successes in acquiring their hardware by one means or another. We could not only test their latest fighters or new radars or missile systems, but actually fly against them. Skunk Works technicians pulled these systems apart, then put them back together, and made tools and spare parts to keep the Russian equipment serviced during testing, so we had a sound notion of what we were up against.

Still, the Air Force had no real interest in using the stealth option to neutralize Soviet defenses. The reason was that while we had learned over the years how to make an airplane less observable to enemy radar, the conventional Pentagon view was that the effectiveness of enemy radar had leaped far ahead of our ability to thwart it. The smart money in aerospace was betting scarce development funds on building airplanes that could avoid the Soviet radar net by coming in just over the treetops, like the new B-1 bomber ordered from Rockwell by the Strategic Air Command, whose purpose was to sneak past ground defenses and deliver a nuclear weapon deep inside the Soviet motherland.

That Pentagon briefing was particularly sobering because it was one of those rare times when our side admitted to a potentially serious gap that tipped the balances against us. I had our advanced planning people noodling all kinds of fantasies — pilotless, remote-controlled drone tactical bombers and hypersonic aircraft that would blister past Soviet radar defenses at better than five times the speed of sound once we solved awesomely difficult technologies. I wish I could claim to have had a sudden two a.m. revelation that made me bolt upright in bed and shout "Eureka!" But most of my dreams involved being chased through a maze of blind alleys by a horde of hostile accountants wielding axes and pitchforks.

The truth is that an exceptional thirty-six-year-old Skunk Works mathematician and radar specialist named Denys Overholser decided to drop by my office one April afternoon and presented me with the Rosetta Stone breakthrough for stealth technology.

The gift he handed to me over a cup of decaf instant coffee would make an attack airplane so difficult to detect that it would be invulnerable against the most advanced radar systems yet invented, and survivable even against the most heavily defended targets in the world.

Denys had discovered this nugget deep inside a long, dense technical paper on radar written by one of Russia's leading experts and published in Moscow nine years earlier. That paper was a sleeper in more ways than one: called "Method of Edge Waves in the Physical Theory of Diffraction," it had only recently been translated by the Air Force Foreign Technology Division from the original Russian language. The author was Pyotr Ufimtsev, chief scientist at the Moscow Institute of Radio Engineering. As Denys admitted, the paper was so obtuse and impenetrable that only a nerd's nerd would have waded through it all — *underlining* yet! The nuggets Denys unearthed were found near the end of its forty pages. As he explained it, Ufimtsev had revisited a century-old set of formulas derived by Scottish physicist James Clerk Maxwell and later refined by the German electromagnetics expert Arnold Johannes Sommerfeld. These calculations predicted the manner in which a given geometric configuration would reflect electromagnetic radiation. Ufimtsev had taken this early work a step further.

"Ben, this guy has shown us how to accurately calculate radar cross sections across the surface of the wing and at the edge of the wing and put together these two calculations for an accurate total."

Denys saw my blank stare. Radar cross section calculations were a branch of medieval alchemy as far as the non-initiated were concerned. Making big objects appear tiny on a radar screen was probably the most complicated, frustrating, and difficult part of modern warplane designing. A radar beam is an electromagnetic field, and the amount of energy reflected back from the target determines its visibility on radar. For example, our B-52, the mainstay long-range bomber of the Strategic Air Command for more than a generation, was the equivalent of a flying dairy barn when viewed from the side on radar. Our F-15 tactical fighter was as big as a two-story Cape Cod house with a carport. It was questionable whether the F-15 or the newer B-70 bomber would be able to survive the ever-improving Soviet defensive net. The F-111 tactical fighter-bomber, using terrain-following radar to fly close to the deck and "hide" in ground clutter, wouldn't survive either. Operating mostly at night, the airplane's radar kept it from hitting mountains, but as we discovered in Vietnam, it also acted like a four-alarm siren to enemy defenses that picked up the F-111 radar from two hundred miles away. We desperately needed new answers, and Ufimtsev had provided us with an "industrial-strength" theory that now made it possible to accurately calculate the lowest possible radar cross section and achieve levels of stealthiness never before imagined.

"Ufimtsev has shown us how to create computer software to accurately calculate the radar cross section of a given configuration, as long as it's in two dimensions," Denys told me. "We can break down an airplane into thousands of flat triangular shapes, add up their individual radar signatures, and get a precise total of the radar cross section."

Why only two dimensions and why only flat plates? Simply because, as Denys later noted, it was 1975 and computers weren't yet sufficiently powerful in storage and memory capacity to allow for three-dimensional designs, or rounded shapes, which demanded enormous numbers of additional calculations. The new generation of supercomputers, which can compute a billion bits of information in a second, is the reason why the B-2 bomber, with its rounded surfaces, was designed entirely by computer computations.

Denys's idea was to compute the radar cross section of an airplane by dividing it into a series of flat triangles. Each triangle had three separate points and required individual calculations for each point by utilizing Ufimtsev's calculations. The result we called "faceting" — creating a three-dimensional airplane design out of a collection of flat sheets or panels, similar to cutting a diamond into sharp-edged slices.

As his boss, I had to show Denys Overholser that I was at least as intellectual and theoretical as Ufimtsev,* so I strummed on my desk importantly and said, "If I understand you, the shape of the airplane would not be too different from the airplane gliders we folded from looseleaf paper and sailed around the classroom behind the teacher's back."

Denys awarded me a "C+" for that try.

The Skunk Works would be the first to try to design an airplane composed entirely of flat, angular surfaces. I tried not to anticipate what some of our crusty old aerodynamicists might say. Denys thought he would need six months to create his computer software based on Ufimtsev's formula. I gave him

---

* Dr. Ufimtsev came to teach electromagnetic theory at UCLA in 1990. Until his arrival here he had remained blissfully unaware of his enormous impact on America's stealth airplane development, but clearly wasn't surprised by the news. "Senior Soviet designers were absolutely uninterested in my theories," he wryly observed.

three months. We code-named the program Echo I. Denys and his old mentor, Bill Schroeder, who had come out of retirement in his eighties to help him after serving as our peerless mathematician and radar specialist for many years, delivered the goods in only five weeks. The game plan was for Denys to design the optimum low observable shape on his computer, then we'd build the model he designed and test his calculations on a radar range.

In those early days of my tenure at the Skunk Works, Kelly Johnson was still coming in twice a week as my consultant as part of his retirement deal. I had mixed feelings about it. On the one hand, Kelly was my mentor and close friend, but it pained me to see so many colleagues crowding into his small office down the hall from mine, taking their work problems to him instead of to me. Of course, I really could not blame them. No one in our shop came close to possessing Kelly's across-the-board technical knowledge, but he didn't just limit himself to providing aerodynamic solutions for stumped engineers; he damned well wanted to know what I was up to, and he wasn't exactly shy about firing off opinions, solicited or not. After a quarter century of working at his side, I knew Kelly's views nearly as well as my own, and I also knew that he would not be thrilled about stealth because he thought the days of manned attack airplanes were definitely numbered. "Goddam it, Ben, the future belongs to missiles. Bombers are as obsolete as the damned stagecoach."

I argued back, "Kelly, the reason they call them missiles, instead of hittles, is that they miss much more than they hit." But Kelly just shook his head.

Several years earlier, we had built a pilotless drone, the D-21, a forty-four-foot manta ray–shaped ramjet that was launched from B-52 bombers to streak high across Communist China and photograph its nuclear missile test facilities. That drone achieved the lowest radar cross section of anything we had ever built in the Skunk Works, and Kelly suggested that we offer our D-21 to the Air Force as a radar-penetrating attack vehicle, with or without a pilot. I put together a small team to begin a modification design, but I couldn't stop thinking about stealth.

That first summer of my takeover, our in-house expert on Soviet weapons systems, Warren Gilmour, attended a meeting at Wright Field, in Ohio, and came back in a dark mood. He marched into my office and closed the door. "Ben, we are getting the shaft in spades," he declared. "One of my friends in the Tactical Air Command spilled the beans. The Defense Department's Advanced Research Projects Agency has invited Northrop, McDonnell Douglas, and three other companies to compete on building a stealthy airplane. They're getting a million bucks each to come up with a proof of concept design, trying to achieve the lowest radar signatures across all the frequencies. If one works, the winner builds two demonstration airplanes. This is right up our alley and we are being locked out in the goddam cold."

This was exactly the kind of project I was looking for. But we had been overlooked by the Pentagon because we hadn't built a fighter aircraft since the Korean War and our track record as builders of low-radar-observable spy planes and drones was so secret that few in the Air Force or in upper-management positions at the Pentagon knew anything about them.

Warren read my mind. "Face it, Ben, those advanced project guys don't have a clue about our spy plane work in the fifties and sixties. I mean, Jesus, if you think racing cars, you think Ferrari. If you think low observables, you must think Skunk Works."

Warren was absolutely right. The trouble was getting permission from our spy plane customer, that legendary sphinx known as the Central Intelligence Agency, to reveal to the Pentagon's competition officials the low observable results we achieved in the 1960s building the Blackbird, which was actually the world's first operational stealth aircraft. It was

140,000 pounds and 108 feet long, about the size of a tactical bomber called the B-58 Hustler, but with the incredibly small radar cross section of a single-engine Piper Cub. In other words, that is what a radar operator would think he was tracking. Its peculiar cobra shape was only part of the stealthy characteristics of this amazing airplane that flew faster than Mach 3 and higher than 80,000 feet. No one knew that its wings, tail, and fuselage were loaded with special composite materials, mostly iron ferrites, that absorbed radar energy rather than returning it to the sender. Basically 65 percent of low radar cross section comes from shaping an airplane; 35 percent from radar-absorbent coatings. The SR-71 was about one hundred times stealthier than the Navy's F-14 Tomcat fighter, built ten years later. But if I knew the CIA, they wouldn't admit that the Blackbird even existed.

Kelly Johnson was regarded almost as a deity at the CIA, and I had him carry our request for disclosure to the director's office. To my amazement, the agency cooperated immediately by supplying all our previously highly classified radar-cross-section test results, which I sent on to Dr. George Heilmeier, the head of DARPA (the Defense Department's Advanced Research Projects Agency), together with a formal request to enter the stealth competition. But Dr. Heilmeier called me, expressing regrets. "Ben, I only wish I had known about this sooner. You're way too late. We've given out all the money to the five competitors." The only possibility, he thought, would be to allow us to enter if we would agree to a one dollar pro forma government contract. As it turned out, if I had done nothing more that first year than refuse that one dollar offer, I had more than earned my salary. I was sitting on a major technological breakthrough, and if I took that government buck, the Feds would own the rights to all our equations, shapes, composites — the works. Lockheed was taking the risks, we deserved the future profits.

It took a lot of arguing at my end, but Dr. Heilmeier finally agreed to let us into the stealth competition with no strings attached, and it was the only time I actually felt good about *not* receiving a government contract. But not Kelly. "You're wasting your time," he told me. "This is like chasing a butterfly in a rain forest because in the end the government won't invest big dollars in stealth, when for the same money they can invest in new missiles."

In part, I think, Kelly was trying to be protective. He didn't want me to risk an embarrassing failure my first turn at bat, pursuing a high-risk project with little apparent long-range potential. I would be spending close to a million dollars of our own development money on this project, and if Kelly was right, I'd wind up with nothing to show for it. Still, I never waivered from believing that stealth could create the biggest Skunk Works bonanza ever. It was a risk well worth taking, proving a technology that could dominate military aviation in the 1980s even more than the U-2 spy plane had impacted the 1950s. At that point the Russians had no satellites or long-range airplanes that could match our missions and overfly us. Stealth would land the Russians on their ear. They had no technology in development that could cope with it. So I resolved to see this project through, even if it meant an early fall from grace. My department heads would go along because they loved high-stakes challenges, with most of the risks falling on the boss. I confided my stealth ambitions to Lockheed's new president, Larry Kitchen, who was himself dancing barefoot on live coals while trying to pull our corporation up to a standing position after the pulverizing year and a half of scandals and bankruptcy. Larry cautioned me: "We need real projects, not pipedreams, Ben. If you've got to take risks, at least make sure you keep it cheap, so I can back you without getting my own head handed to me. And if something goes sour, I want to be the first to know. My blessings."

Good man, Larry Kitchen. After all, he had also approved hiring me as Kelly's successor.

Denys Overholser reported back to me on May 5, 1975, on his attempts to design the stealthiest shape for the competition. He was wearing a confident smile as he sat down on the couch in my office with a preliminary designer named Dick Scherrer, who had helped him sketch out the ultimate stealth shape that would result in the lowest radar observability from every angle. What emerged was a diamond beveled in four directions, creating in essence four triangles. Viewed from above the design closely resembled an Indian arrowhead.

Denys was a hearty outdoorsman, a cross-country ski addict and avid mountain biker, a terrific fellow generally, but inexplicably fascinated by radomes and radar. That was his specialty, designing radomes — the jet's nose cone made out of noninterfering composites, housing its radar tracking system. It was an obscure, arcane specialty, and Denys was the best there was. He loved solving radar problems the way that some people love crossword puzzles.

"Boss," he said, handing me the diamond-shaped sketch, "Meet the Hopeless Diamond."

"How good are your radar-cross-section numbers on this one?" I asked.

"Pretty good." Denys grinned impishly. "Ask me, 'How good?' "

I asked him and he told me. "This shape is one thousand times less visible than the least visible shape previously produced at the Skunk Works."

"Whoa!" I exclaimed. "Are you telling me that this shape is a thousand times less visible than the D-21 drone?"

"You've got it!" Denys exclaimed.

"If we made this shape into a full-size tactical fighter, what would be its equivalent radar signature . . . as big as what — a Piper Cub, a T-38 trainer . . . what?"

Denys shook his head vigorously. "Ben, understand, we are

talking about a *major*, *major*, big-time revolution here. We are talking *infinitesimal*."

"Well," I persisted, "what does that mean? On a radar screen it would appear as a . . . what? As big as a condor, an eagle, an owl, a what?"

"Ben," he replied with a loud guffaw, "try as big as an eagle's *eyeball*."

2

ENGINES BY GE, BODY BY HOUDINI

KELLY JOHNSON was not impressed. He took one look at Dick Scherrer's sketch of the Hopeless Diamond and charged into my office. Unfortunately, he caught me leaning over a work table studying a blueprint, and I never heard him coming. Kelly kicked me in the butt — hard too. Then he crumpled up the stealth proposal and threw it at my feet. "Ben Rich, you dumb shit," he stormed, "have you lost your goddam mind? This crap will never get off the ground."

Frankly, I had the feeling that there were a lot of old-timers around the Skunk Works who wanted badly to do what Kelly had just done. Instead they did it verbally and behind my back. These were some our most senior aerodynamicists, thermodynamicists, propulsion specialists, stress and structures and weight engineers, who had been building airplanes from the time I was in college. They had at least twenty airplanes under their belts and were walking aviation encyclopedias and living parts catalogs. Over the years they had solved every conceivable problem in their specialty areas and damned well knew what worked and what didn't. They were crusty and stiff-necked at times, but they were all dedicated, can-do guys who worked fourteen-hour days seven days a week for months on end to make a deadline. Self-assurance came from experiencing many more victories than defeats. At the Skunk Works we designed practical, used off-the-shelf parts whenever possible, and did things right the first time. My wing man, for example, had designed twenty-seven wings on previous Skunk Works' airplanes before tackling the Hopeless Diamond. All of us had been trained by Kelly Johnson and believed fanatically in his insistence that an airplane that looked beautiful would fly the same way. No one would dare to claim that the Hopeless Diamond would be a beautiful airplane. As a flying machine it looked *alien*.

Dave Robertson, one of Kelly's original recruits and aerospace's most intuitively smart hydraulic specialist, ridiculed our design by calling it "a flying engagement ring." Dave seldom minced words; he kept a fourteen-inch blowgun he had fashioned out of a jet's tailpipe on his desk and would fire clay pellets at the necks of any other designers in the big drafting room who got on his nerves. Robertson hated having anyone look over his shoulder at his drawing and reacted by grabbing a culprit's tie and cutting it off with scissors. Another opponent was Ed Martin, who thought that anyone who hadn't been building airplanes since the propeller-driven days wasn't worth talking to, much less listening to. He called the Hopeless Diamond "Rich's Folly." Some said that Ed's bark was worse than his bite, but those were guys who didn't know him.

Most of our veterans used slide rules that were older than Denys Overholser, and they wondered why in hell this young whippersnapper was suddenly perched on a throne as my guru, seemingly calling the shots on the first major project under my new and untested administration. I tried to explain that stealth technology was in an embryonic state and barely understood until Denys unearthed the Ufimtsev theory for us; they remained unconvinced even when I reminded them that until Denys had come along with his revelation, we had known only two possibilities to reduce an airplane's radar detection. One way was to coat the fuselage, tail, and wing surfaces with special composite materials that would absorb incoming electromagnetic energy from radar waves instead of bouncing it back to the sender. The other method was to

construct an airplane out of transparent materials so that
the radar signals would pass through it. We tried an experimental transparent airplane back in the early 1960s and to
our dismay discovered that the engine loomed ten times bigger on radar than the airplane because there was no way to
hide it.

So all of us, myself especially, had to trust that Denys Overholser, with his boyish grin and quiet self-confidence, *really*
knew what in hell he was talking about and could produce
big-time results. Dick Cantrell, head of our aerodynamics
group, suggested burning Denys at the stake as a heretic and
then going on to conventional projects. Cantrell, normally as
soft-spoken and calm as Gregory Peck, whom he vaguely resembled, nevertheless had the temperament of a fiery Sa-
vonarola when, as in this instance, basics of fundamental
aerodynamics were tossed aside in deference to a new technology understood only by witches and mathematical
gnomes. But after a couple of hours of listening to Overholser's explanations of stealth, Dick dropped his lanky frame
onto the chair across from my desk and heaved a big sigh.
"Okay, Ben," he muttered, "I surrender. If that flat plate concept is really as revolutionary as that kid claims in terms of
radar cross section, I don't care what in hell it looks like, I'll
get that ugly son-of-a-bitch to fly."

We could get the Statue of Liberty to do barrel rolls with the
onboard computers that achieved aerodynamic capability by
executing thousands of tiny electrohydraulic adjustments every second to an airplane's control surfaces. This computerized enhanced flight stability gave us latitude in designing
small, stealthy wings and short tails and mini-wing flaps, and
left the awesome problems of unstable pitch and yaw to the
computers to straighten out. Without those onboard computers, which the pilots called "fly-by-wire," since electric wiring now replaced conventional mechanical control rods, our
diamond would have been hopeless indeed. But even with the

powerful onboard computers, getting into the sky, as Kelly's boot to my butt suggested, would be far from a cakewalk.

We had a very strong and innovative design organization of about a dozen truly brilliant engineers, working at their drawing boards in a big barnlike room on the second floor of our headquarters building, who simply could not be conned or browbeaten into doing anything they knew would not work. One day, Kelly called upstairs for an engineer named Bob Allen. "Bob Allen there?" he asked. Whoever answered the phone replied, "Yeah, he is." And hung up. Kelly was livid, but deep down he appreciated the fiesty independence of his best people. The designers were either structural specialists who planned the airframe or systems designers who detailed the fuel, hydraulics, electrical, avionics, and weapons systems. In many ways they comprised the heart and soul of the Skunk Works and also were the most challenged by the structural demands of the new stealth technology. Thanks to Ufimtsev's breakthrough formula, they were being told to shape an airplane entirely with flat surfaces and then tilt the individual panels so that radar energy scattered away and not back to the source. The airplane would be so deficient in lift-drag ratio that it would probably need a computer the size of Delaware to get it stable and keep it flying.

Several of our aerodynamics experts, including Dick Cantrell, seriously thought that maybe we would do better trying to build an actual flying saucer. The shape itself was the ultimate in low observability. The problem was finding ways to make a saucer fly. Unlike our plates, it would have to be rotated and spun. But how? The Martians wouldn't tell us.

During those early months of the Hopeless Diamond, I dug in my heels. I forced our in-house doubters to sit down with Denys and receive a crash course on Stealth 101. That helped to improve their confidence quotient somewhat, and although I acted as square-shouldered as Harry Truman challenging the Republican Congress, deep down I was suffering bouts of angst myself, wondering if Kelly and some of the other skeptics had it right while I was being delusional. I kept telling myself that the financial and personal risks in pursuing this project were minimal compared to its enormous military and financial potential. But the politics of the situation had me worried: stealth would have been a perfect third project for me, after two reassuring successes under my belt.

But if stealth failed, I could hear several of my corporate bosses grousing: "What's with Rich? Is he some sort of flake? Kelly would never have undertaken such a dubious project. We need to take charge of that damned Skunk Works and make it practical and profitable again."

Kelly Johnson would never double-cross me by bad-mouthing the stealth project in the corridors of the Skunk Works, but all of us knew Kelly too well not to be able to read his mood and mind. If he didn't like something or someone, it was as obvious as a purple pimple on the tip of his nose. So I had him in for lunch and said, "Look, Kelly, I know you find this design aesthetically offensive, but I want you to do me one favor. Sit down with this guy, Overholser, and let him answer your questions about stealth. He's convinced me that we are onto something enormously important. Kelly, this diamond is somewhere between ten thousand and one hundred thousand times lower in radar cross section than any U.S. military airplane or any new Russian MiG. Ten thousand to one hundred thousand times, Kelly. Think of it!"

Kelly remained unmoved. "Theoretical claptrap, Ben. I'll bet you a quarter that our old D-21 drone has a lower cross section than that goddam diamond."

We had a ten-foot wooden model of the diamond, and we took it and the original wooden model for the manta ray–shaped D-21 drone and put them side by side into an electromagnetic chamber and cranked up the juice.

That date was September 14, 1975, a date etched forever in my memory because it was about the only time I ever won a quarter from Kelly Johnson. I had lost about ten bucks' worth of quarters to him over the years betting on technical matters. Like me, my colleagues collected quarters from Kelly just about as often as they beat him at arm wrestling. He had been a hod carrier as a kid and had arms like ship's cables. He once sprained the wrist of one of our test pilots so badly he put the poor guy out of action for a month. So winning a quarter was a very big deal, in some ways even more satisfying than winning the Irish Sweepstakes. (Depending on the size of the purse, of course.)

I really wanted a photographer around for historical purposes to capture the expression on Kelly's big, brooding moon-shaped mug when I showed him the electromagnetic chamber results. Hopeless Diamond was exactly as Denys had predicted: a thousand times stealthier than the twelve-year-old drone. The fact that the test results matched Denys's computer calculations was the first proof that we actually knew what in hell we were doing. Still, Kelly reacted about as graciously as a cop realizing he had collared the wrong suspect. He grudgingly flipped me the quarter and said, "Don't spend it until you see the damned thing fly."

But then he sent for Denys Overholser and grilled the poor guy past the point of well-done on the whys and hows of stealth technology. He told me later that he was surprised to learn that with flat surfaces the amount of radar energy returning to the sender is independent of the target's size. A small airplane, a bomber, an aircraft carrier, all with the same shape, will have identical radar cross sections. "By God, I never would have believed that," he confessed. I had the feeling that maybe he still didn't.

Our next big hurdle was to test a ten-foot wooden model of the Hopeless Diamond on an outdoor radar test range near Palmdale, on the Mojave desert. The range belonged to McDonnell Douglas, which was like Buick borrowing Ford's test track to road test an advanced new sports car design, but I had no choice since Lockheed didn't own a radar range. Our model was mounted on a 12-foot-high pole, and the radar dish zeroed in from about 1,500 feet away. I was standing next to the radar operator in the control room. "Mr. Rich, please check on your model. It must've fallen off the pole," he said. I looked. "You're nuts," I replied. "The model is up there." Just then a black bird landed right on top of the Hopeless Diamond. The radar operator smiled and nodded. "Right, I've got it now." I wasn't about to tell him he was zapping a crow. His radar wasn't picking up our model at all.

For the first time, I felt reassured that we had caught the perfect wave at the crest and were in for one terrifically exciting ride. I saw firsthand how invisible that diamond shape really was. So I crossed my fingers and said a silent prayer for success in the tests to follow.

Other Voices
Denys Overholser

In October 1975, Ben Rich informed me that we and Northrop had won the first phase of the competition and would now contest against each other's designs in a high noon shoot-out at the Air Force's radar test range in White Sands, New Mexico. The two companies were each given a million and a half dollars to refine the models and told to be ready to test in four months.

The government demanded competition on any project, but that Hopeless Diamond shape was tough to beat. We built the model out of wood, all flat panels, thirty-eight feet, painted black. And in March 1976 we hauled it by truck to New Mexico. The White Sands radar range was used to test unarmed nuclear warheads, and their radars were the most sensitive and powerful in the free world.

The tests lasted a month. I never did see the Northrop model because under the ground rules we tested separately, on different days. In the end we creamed them. Our diamond was ten times less visible than their model. We achieved the lowest radar cross sections ever measured. And the radar range test results precisely matched the predictions of our computer software. This meant we could now confidently predict radar cross section for any proposed shape, a unique capability at that point in time.

The range was as flat as a tabletop; the pole downrange was in a direct line with five different radar antenna dishes, each targeting a different series of frequencies. The model was mounted atop the pylon and then rotated in front of the radar beam. Well, two very funny things happened. The first day we placed our model on the pole, the pole registered many times brighter than the model. The technicians had a fit. They had thought their poles were invisible, but the trouble was nobody had ever built a model that was so low in radar signature to show them how wrong they really were. Their pole registered minus 20 decibels — okay as long as the model on top was greater than 20. But when the model was registering an unheard-of lower value, the pole intruded on the testing. An Air Force colonel confronted me in a fit of pique: "Well," he snorted, "since you're so damned clever, build us a new pole." I thought, Oh, sure. Build a tower that's ten decibels lower than the model. Lots of luck.

In the end we had to team up with Northrop to pay for the poles, because the Air Force wasn't about to foot the bill. It cost around half a million dollars. And I designed a double-wedge pylon which they tested on a 50,000-watt megatron, state of the art in transmitters, that could pick up an object the size of an ant from a mile away. On that radar the pole was about the size of a bumblebee. John Cashen, who was Northrop's stealth engineer, was in the control room when they fired up the radar. And I overheard their program manager whisper to John: "Jesus, if they can do that with a frigging pole, what can they do with their damned model?"

Ben called me every day for the latest results. The model

was measuring approximately the equivalent of a golf ball. One
morning we counted twelve birds sitting on the model on top
of the pole. Their droppings increased the radar cross section
by one and a half decibels. Three decibels is the equivalent of
doubling its cross section. And as the day heated on the desert,
inversion layers sometimes bent the radar off the target. One
day, while using supersensitive radar, the inversion layer bent
the beam off the target, making us four decibels better than we
deserved. I saw that error, but the technician didn't. What the
hell, it wasn't my job to tell him he had a false pattern. I figured
Northrop probably benefited from a few of them too, and it
would all come out in the wash.

But then Ben Rich called me and said, "Listen, take the best
pattern we've got, calculate the cross section level, and tell me
the size of the ball bearing that matches our model." This was a
Ben Rich kind of idea. The model was now shrunk down from
a golf ball to a marble because of bad data. But it was official
bad data, and no one knew it was bad except little me.

So Ben went out and bought ball bearings and flew to the
Pentagon and visited with the generals and rolled ball bearings
across their desktops and announced, "Here's your airplane!"
Those generals' eyes bugged out of their heads. John Chasen
was livid when he found out about it because he hadn't thought
of it first. "That goddam Ben Rich," he fumed. And a few
months later, Ben had to stop rolling them across the desk of
anyone who wasn't cleared.

In early April 1976, I got the word that we had officially won
the competition with Northrop and would go on to build two
experimental airplanes based on our Hopeless Diamond design. The program was now designated under the code name
Have Blue. We knew we could produce a model with spectacularly low radar signatures, but the big question was
whether we could actually build an airplane that would enjoy
the same degree of stealthiness. A real airplane was not only

much larger, but also loaded with all kinds of anti-stealth features — a cockpit, engines, air scoops and exhausts, wing and tail flaps, and landing gear doors. In any airplane project the design structures people, the aerodynamics group, and the propulsion and weight specialists all argue and vie for their points of view. In this case, however, I served notice that Denys Overholser's radar cross section group had top priority. I didn't give a damn about the airplane's performance characteristics because its only purpose was to demonstrate the lowest radar signature ever recorded. I joked that if we couldn't get her airborne, maybe we could sell her as a piece of modern art sculpture.

I assigned the design project to Ed Baldwin, who was our best and most experienced structural engineer. "Baldy" had started out with Kelly designing the P-80, America's first jet fighter, in 1945, and had designed the configuration of the U-2 spy plane. His task was to take the preliminary design concept of the Hopeless Diamond and make it practical so that it could actually fly. Dick Scherrer had done the preliminary design, laying out the basic shape, and Baldwin had to make certain that the shape's structure was sound and practical; he would determine its radius, its thicknesses, its ability to withstand certain loads, the number of parts. "Baldy" would put the rubber on the ramp.

All of our structure and wing guys worked for him, and Baldwin enjoyed badgering aerodynamicists, especially in meetings where he could score points with his fellow designers by making aerodynamicists squirm or turn beet red in fury. One on one, Baldy was a pleasant chap — at least moderately so for a crusty Skunk Works veteran — but in meetings we were all fair game and he was a bad-tempered grizzly. Early on, for example, he got into a heated exchange with a very proper Britisher named Alan Brown, our propulsion and stealth expert, about some aspect of the structure he was designing. Baldwin turned crimson. "Goddam it, Brown," he said, "I'll design this friggin' airplane and you can put on the friggin' stealth afterwards."

The airplane Baldy designed was a single-seat, twin-engined aircraft, 38 feet long, with a wingspan of 22 feet and a height of slightly more than seven feet. Its gross weight was 12,000 pounds. The leading edge of the delta wing was razor-sharp and swept back more than 70 degrees. To maintain low infrared signatures, the airplane could not go supersonic or have an afterburner because speed produced surface heating — acting like a spotlight for infrared detection. Nor did we want the airplane to be aurally detected from the ground. For acoustical reasons we had to make sure we had minimized engine and exhaust noise by using absorbers and shields. To keep it from being spotted in the sky, we decided to use special additives to avoid creating exhaust contrails. And to eliminate telltale electromagnetic emissions, there was no internal radar system on board.

Our airplane wasn't totally invisible, but it held the promise of being so hard to detect that even the best Soviet defenses could not accomplish a fatal lock-on missile cycle in time to thwart its mission. If they could detect a fighter from a hundred miles out, that airplane was heading for the loss column. The long-range radar had plenty of time to hand off the incoming intruder to surface-to-air missile batteries, which, in turn, would fire the missiles and destroy it. Early-warning radar systems could certainly see us, but not in time to hand us over to missile defenses. If the first detection of our airplane was at fifteen miles from target, rather than at fifty miles, there simply would be no time to nail us before we hit the target. And because we were so difficult to detect, even at fifteen miles, radar operators would also be thwarted while trying to detect us through a confusing maze of ground clutter.

I had asked Kelly to estimate the cost of building these two experimental Have Blue airplanes. He came back with the figure of $28 million, which turned out to be almost exactly right. I asked the Air Force for $30 million, but they had only $20 million to spend in discretionary funds for secret projects by which they bypassed congressional appropriations procedures. So, in the late spring of 1976, I was forced to go begging for the missing $10 million to our CEO, Bob Haack, who was sympathetic but not particularly enthusiastic. He said, "Look, Ben, we're in tough straits right now. I don't think we can really afford this." I pushed a little harder and got him to agree to let me present the proposal to the full board of directors. Bob set up the meeting, and I just laid it all out. Larry Kitchen, Lockheed's president, and Roy Anderson, the vice chairman, spoke up enthusiastically in support. I told the board I thought we were dealing with a project that had the potential for $2 to $3 billion in future sales. I predicted we would be building stealth fighters, stealth missiles, stealth ships, the works. I was accused of hyperbole by one or two directors, but in the end I got my funding, and as time went on my sales predictions proved to be conservatively low.

Even worse, I began picking up rumors that certain officials at the Pentagon were accusing me of rigging the test results of the radar range competition against Northrop. An Air Force general called me, snarling like a pit bull. "Rich, I'm told you guys are pulling a fast one on us with phony data." I was so enraged that I hung up on that son of a bitch. No one would have ever dared to accuse Kelly Johnson's Skunk Works of rigging any data, and by God, no one was going to make that accusation against Ben Rich's operation either. Our integrity was as important to all of us as our inventiveness. The accusation, I discovered, was made by a civilian radar expert advising the Air Force, who had close ties to leading manufacturers of electronic jamming devices installed in all Air Force planes to fool or thwart enemy radar and missiles. If stealth was as good as we claimed, those companies might be looking for new work.

His motivation for bad-mouthing us was obvious; but it was

equally apparent that we were unfairly being attacked without any effective way for me to defend the Skunk Works' integrity from three thousand miles away. So I invited one of the nation's most respected radar experts to Burbank to personally test and evaluate our stealth data. MIT's Lindsay Anderson accepted my invitation in the late summer of 1976 and arrived at my doorstep carrying a bag of ball bearings in his briefcase. The ball bearings ranged in size from a golf ball to an eighth of an inch in diameter. Professor Anderson requested that we glue each of these balls onto the nose of the Hopeless Diamond and then zap them with radar. This way he could determine whether our diamond had a lower cross section than the ball bearings. If the diamond in the background proved to be brighter than the ball in the foreground, then the ball could not be measured at all. That got me a little nervous because nothing should measure less than an eighth-of-an-inch ball bearing, but we went ahead anyway. As it turned out, we measured all the balls easily — even the eighth-of-an-inch one — and when Professor Anderson saw that the data matched the theoretical calculated value of the ball bearings, he was satisfied that all our claims were true.

That was the turning point for the entire stealth adventure, which could have ended right there if Lindsay Anderson had reinforced the accusation that we were being unscrupulous and presenting bogus data. But once he corroborated our achievement back in Washington, I was informed by a telegram from the Air Force chief of staff that Have Blue was now classified "Top Secret — Special Access Required." That security classification was rare — clamped only on such sensitive programs as the Manhattan Project, which created the first atomic bomb during World War II. My first reaction was "Hooray, they finally realize how significant this technology really is," but Kelly set me straight and with a scowl urged me to cancel the whole damned project right then and there.

"Ben," Kelly warned me, "the security they're sticking onto this thing will kill you. It will increase your costs twenty-five percent and lower your efficiency to the point where you won't get any work done. The restrictions will eat you alive. Make them reclassify this thing or drop it." On matters like that, Kelly was seldom wrong.

Other Voices

General Larry D. Welch
(Air Force Chief of Staff from 1986 to 1990)

In 1976, I was a brigadier general in charge of planning at the Tactical Air Command at Langley, Virginia, when my boss, General Bob Dixon, called me one afternoon and told me to drop whatever I was doing to attend an extremely classified briefing. He said, "The only people I've cleared for this briefing are you and one other general officer." I went over to headquarters and discovered that Ben Rich of Lockheed's Skunk Works was making a presentation about producing an operational stealth aircraft. Bill Perry, who ran R & D at the Pentagon, had sent him over to us because Dr. Perry was very interested in the stealth concept and wanted our input. Ben spoke only about twenty minutes. After he left, we went into General Dixon's office and he asked, "Well, what do you two think?" I said, "Well, sir, from a purely technical standpoint I don't have a clue about whether this concept is really achievable. Frankly, I'm not even sure the goddam thing will fly. But if Ben Rich and the Skunk Works say that they can deliver the goods, I think we'd be idiots not to go along with them." General Dixon wholeheartedly agreed with me. And so we started the stealth program on the basis of Ben's twenty-minute presentation and a hell of a lot of faith in Ben Rich & Company. And that faith was based on long personal experience.

Way back when I was a young colonel working in the fighter

division — this would be the early seventies — I was tasked to come up with a realistic cost estimate for a revolutionary tactical fighter with movable wings called the FX, which later became the F-15. Inside the Air Force there was a lot of controversy about costs that ranged from $3.5 million to $8.5 million. Before we could ask Congress for money, we had to reach some sort of consensus, so I persuaded my boss to let me go out to the Skunk Works in Burbank and get their analysis because they were the best in the business. So I flew out and sat down with Kelly Johnson and Ben Rich. After drinking exactly one ounce of whisky from one of Kelly's titanium shot glasses, we got down to business. Ben and Kelly worked out the figures on a piece of paper — Okay, here's what the avionics will cost, and the airframe, and so on. The overall cost they predicted per airplane would be $7 million. And so we went to Congress and told them that the FX would cost between $5 million and $7 million. The day we delivered that airplane the cost came out to $6.8 million per airplane in 1971 dollars.

So I had supreme confidence that Ben and his people would deliver superbly on stealth. There were only five of us at headquarters cleared for the stealth program, and I became the head logistician, the chief operations officer, and the civil engineer for the Air Force side. The management approach we evolved was unique and marvelous. Once a month, I'd meet with Dr. Perry at the Pentagon and inform him about decisions we required from him as Under Secretary of Defense. Sometimes he agreed, sometimes not, but we never had delays or time wasted with goddam useless meetings. Because we were so highly classified, the bureaucracy was cut out and that made a tremendous difference. Frankly, that was a damned gutsy way to operate inside the Pentagon, but the reason we could afford to be so gutsy was our abiding faith in the Skunk Works.

Before the government would sign a contract with me I had to submit for approval a security plan, detailing how we would tighten all the hatches of what was already one of the most secure operations in the defense industry. Hell, we already operated without windows and behind thick, eavesdrop-proof walls. We had special bank-vault conference rooms, lined with lead and steel, for very sensitive discussions about very secret matters. Still, the Air Force required me to change our entire security system, imposing the kinds of strictures and regulations that would drive us all nuts in either the short or long run. Every piece of paper dealing with the project had to be stamped top secret, indexed in a special security filing system, and locked away. Full field investigations were demanded of every worker having access to the airplane. They imposed a strictly enforced two-man rule: no engineer or shop worker could be left alone in a room with a blueprint. If one machinist had to go to the toilet, the co-worker had to lock up the blueprint until his colleague returned.

Only five of us were cleared for top secret and above, and over the years we had worked on tremendously sensitive projects without ever suffering a leak or any known losses to espionage. In fact, Kelly evolved his own unorthodox security methods, which worked beautifully in the early days of the 1950s. We never stamped a security classification on any paperwork. That way, nobody was curious to read it. We just made damned sure that all sensitive papers stayed inside the Skunk Works.

My biggest worry was clearing our workers for this project. They needed Special Access clearances, and I had to make the case for their Need to Know on an individual basis. But the government, not the employer, was the final arbiter of who was granted or denied access. The Air Force security people made the decision and offered no explanation about why certain of my employees were denied access to the program. No one in Washington conferred with me or asked my opinion or sought my advice. I knew my people very well. Some were horse players, several were skirt chasers, a few were not always prompt about paying their bills. For all I knew some of my best people might be part-time transvestites. I had no doubt that some of the younger ones may have indulged in "recreational drugs," like toking marijuana at rock shows. Any of these "sins" could sink a valuable worker. I did win a couple of important concessions: the Air Force agreed that only those few technicians with a need to know the airplane's radar cross section would require the complete full field investigation, which took around nine months, and I was granted temporary clearances for twenty specialists working on particular sensitive aspects of Have Blue. Most important, I raised so much cain that Air Force security finally granted me a "grandfather clause" for many of our old-timers who had been working on all our secret projects since the days of the U-2. They were granted waivers to work on Have Blue.

But security's dragnet poked and prodded into every nook and cranny of our operation. Keith Beswick, head of our flight test operations, designed a coffee mug for his crew with a clever logo showing the nose of Have Blue peeking from one end of a big cloud with a skunk's tail sticking out the back end. Because of the picture of the airplane's nose, security classified the mugs as top secret. Beswick and his people had to lock them away in a safe between coffee breaks. The airplane itself had to be stamped SECRET on the inside cockpit door. I was named its official custodian and had to sign for it whenever it left its hangar area and was test-flown. If it crashed, I was personally responsible for collecting every single piece of it and turning all of it over to the proper authorities.

These draconian measures hobbled us severely at times, tested my patience beyond endurance, and gave Kelly every right to scold, "Goddam it, Rich, I told you so." At one point I had to memorize the combinations to three different security safes just to get work done on a daily basis. A few guys with lousy memories tried to cheat and carried the combination numbers in their wallets. If security caught them, they could be fired. Security would snoop in our desks at night to search for classified documents not locked away. It was like working at KGB headquarters in Moscow.

The Air Force wanted the two test planes in only fourteen months. Over the years we had developed the concept of using existing hardware developed and paid for by other programs to save time and money and reduce the risks of failures in prototype projects. I worked an agreement with the Air Force to supply me with the airplane engines. They assigned an expediter named Jack Twigg, a major in the Tactical Air Command, who was cunning and smart. Jack requisitioned six engines from the Navy. He went to General Electric's jet engine division, did some fast talking to the president and plant manager, got some key people to look the other way while he carted away the six J-85 engines we needed right off their assembly line, and had them shipped in roundabout ways, so that nobody knew the Skunk Works was the final destination. We put two engines in each experimental airplane and had a couple of spares. Jack was a natural at playing James Bond: he ordered parts in different batches and had them shipped using false return addresses and drop boxes.

We begged and borrowed whatever parts we could get our hands on. Since this was just an experimental stealth test vehicle destined to be junked at the end, it was put together with avionics right off the aviation version of the Kmart shelf: we took our flight control actuators from the F-111 tactical bomber, our flight control computer from the F-16 fighter, and the inertial navigation system from the B-52 bomber. We took the servomechanisms from the F-15 and F-111 and modified them, and the pilot's seat from the F-16. The heads-up display was designed for the F-18 fighter and adapted for our airplane. In all we got about $3 million worth of equipment from the Air Force. That was how we could build two airplanes and test them for two years at a cost of only $30 million. Normally, a prototype for an advanced technology airplane would cost the government three or four times as much.

Only the flight control system was specially designed for Have Blue, since our biggest sweat was aerodynamics. We decided to use the onboard computer system of General Dynamics's small-wing lightweight fighter, the F-16, which was designed unstable in pitch; our airplane would be unstable in all three axes — a dubious first that brought us plenty of sleepless nights. But we had our very own Bob Loschke, acknowledged as one of the very best onboard computer experts in aerospace, to adapt the F-16's computer program to our needs. We flew the airplane avionically on the simulator flight control system and kept modifying the system to increase stability. It was amazing what Loschke could accomplish artificially by preempting the airplane's unstable responses and correcting them through high-powered computers.

The pilot tells the flight control system what he wants it to do just by normal flying: maneuvering the throttle and foot pedals directing the control surfaces. The electronics will move the surfaces the way the pilot commands, but often the system will automatically override him and do whatever it has to do to keep the system on track and stable without the pilot even being aware of it. Our airplane was a triumph of computer technology. Without it, we could not even taxi straight.

In July 1976, we began building the first of two Have Blue prototypes in Building 82, one of our big assembly hangars, the size of three football fields. We had our own very unique method for building an airplane. Our organizational chart consisted of an engineering branch, a manufacturing branch, an inspection and quality assurance branch, and a flight testing branch. Engineering designed and developed the Have Blue aircraft and turned it over to the shop to build. Our engineers were expected on the shop floor the moment their blueprints were approved. Designers lived with their designs through fabrication, assembly, and testing. Engineers couldn't just throw their drawings at the shop people on a take-it-or-leave-it basis and walk away.

Our senior shop people were tough, experienced SOBs and not shy about confronting a designer on a particular drawing and letting him know why it wouldn't work. Our designers spent at least a third of their day right on the shop floor; at the same time, there were usually two or three shop workers up in the design room conferring on a particular problem. That was how we kept everybody involved and integrated on a project. My weights man talked to my structures man, and my structures man talked to my designer, and my designer conferred with my flight test guy, and they all sat two feet apart, conferring and kibitzing every step of the way. We trusted our people and gave them the kind of authority that was unique in aerospace manufacturing. Above all, I didn't second-guess them.

Our manufacturing group consisted of the machine shop people, sheet metal fabrication and assemblers, planners, tool designers, and builders. Each airplane required its own special tools and parts, and in projects like Have Blue, where only two prototypes were involved, we designed and used wooden tools to save time and money. When the project ended, we just threw them away.

The shop manufactured and assembled the airplane, and the inspection and quality assurance branch checked the product at all stages of development. That was also unique with us, I think. In most companies quality control reported to the head of the shop. At the Skunk Works quality control reported directly to me. They were a check and balance on the work of the shop. Our inspectors stayed right on the floor with the machinists and fabricators, and quality control inspections occurred almost daily, instead of once, at the end of a procedure. Constant inspection forced our workers to be super-critical of their work before passing it on. Self-checking was a

Skunk Works concept now in wide use in Japanese industry and called by them Total Quality Management.

Our workers were all specialists in specific sections of the airplane: fuselage, tail, wings, control surfaces, and power plant. Each section was built separately then brought together and assembled like a giant Tinkertoy. We used about eighty shop people on this project, and because we were in a rush and the airplane was small, we stood it on its tail and assembled it vertically. That way, the assemblers could work on the flat, plated structural frame, front and back, asses to elbows, simultaneously. I kept Alan Brown, our stealth engineer, on the floor all the time to answer workers' questions.

Flat plates, we discovered, were much harder to tool than the usual rounded surfaces. The plates had to be absolutely perfect to fit precisely. We also had nagging technical headaches applying the special radar-absorbing coatings to the surfaces. Each workday the problems piled higher and I sat behind Kelly's old desk reaching for my industrial-size bottle of headache tablets. Meanwhile, the Navy came to us to test the feasibility for a stealthy weapons system and set up their own top secret security system that was twice as stringent as the Air Force's. We had to install special alarm systems that cost us a fortune in the section of our headquarters building devoted to the naval work. And we were also doing some prototype work for the Army on stealthy munitions.

In the midst of all this interservice rivalry, security, and hustle and bustle, Major General Bobby Bond, who was in charge of tactical air warfare, came thundering into the Skunk Works with blood in his eye on a boiling September morning. The Santa Ana winds were howling and half of L.A. was under a thick pall of smoke from giant brush fires, mostly started by maniacs with matches. My asthma was acting up and I had a lousy headache and I was in no mood for a visit from the good general, even though I had a special regard for the guy. But General Bond was a brooder and a worrier, who drove me and everyone else absolutely bonkers at times. He always thought he was being shortchanged or victimized in some way. He pounded on my desk and accused me of taking some of my best workers off his Have Blue airplane to work on some rumored secret Navy project. I did my best to look hurt and appeased Bobby by even raising my right hand in a solemn oath. I told myself, So, it's a little white lie. What else can I do? The Navy project is top secret and Bond has no need to know. We could both go to jail if I told him what was really up.

Unfortunately, on the way out to lunch, the general spotted a special lock and alarm system above an unmarked door which he knew from prowling the rings of the Pentagon was used only by the Navy on its top secret projects. Bond squeezed my arm. "What's going on inside that door?" he demanded to know. Before I could think up another lie, he commanded me to open up that door. I told him I couldn't; he wasn't cleared to peek inside. "Rich, you devious bastard, I'm giving you a direct order, open up that goddam door this instant or I'll smash it down myself with a goddam fire ax." The guy meant every word. He began pounding on the door until it finally opened a crack, and he forced his way in. There sat a few startled Navy commanders.

"Bobby, it isn't what you think," I lied in vain.

"The hell it isn't, you lying SOB," he fumed.

I surrendered, but not gracefully. I said, "Okay, you got me. But before we go to lunch you're going to have to sign an inadvertent disclosure form or security will have both our asses." The Navy, of course, was outraged at both of us. An Air Force general seeing their secret project was as bad as giving a blueprint to the Russians.

Bobby* didn't worry about the Navy very long, because we gave him far bigger worries than that: four months before we

---

* General Bond was later killed in a test flight. Because of that tragedy the Pentagon ruled that general officers could no longer do flight tests.

were supposed to test-fly Have Blue our shop mechanics went out on strike.

The International Association of Machinists' negotiations with the Lockheed corporation on a new two-year contract failed in late August 1977. Our workers hit the bricks just as Have Blue was going into final assembly, perched on its jig with no hydraulic system, no fuel system, no electronics or landing gear. There seemed to be no way we would be ready to fly by December 1, our target date, and our bean counters wanted to inform the Air Force brass that we would be delayed one day for each day of the strike. But Bob Murphy, our veteran shop superintendent, insisted that he could get the job done on time and meet our commitment for first flight. To Murphy, it was a matter of stubborn Skunk Works pride.

Bob put together a shop crew of thirty-five managers and engineers who worked twelve hours a day, seven days a week, over the next two months. Fortunately, most of our designers were all great tinkerers, which is probably why they were drawn to engineering in the first place. Murphy had Beswick, our flight test head, working with a shop supervisor named Dick Madison assembling the landing gear. Murphy himself put in the ejection seat and flight controls; another shop supervisor named John Stanley worked alone on the fuel system. Gradually, the airplane began coming together, so that by early November Have Blue underwent strain gauge calibrations and fuel system checkout. Because Have Blue was about the most classified project in the free world, it couldn't be rolled outdoors, so the guys defied rules and regulations and ran fuel lines underneath the hangar doors to tank up the airplane and test for leaks. But how could we run engine tests?

Murphy figured out a way. He rolled out the plane after dark to a nearby blast fence about three hundred yards from the Burbank Airport main runway. On either side he placed two tractor trailer vans and hung off one end a large sheet of canvas. It was a gerry-built open-ended hangar that shielded Have Blue from view; security approved provided we had the airplane in the hangar before dawn.

Meanwhile an independent engineering review team, composed entirely of civil servants from Wright Field in Ohio, flew to Burbank to inspect and evaluate our entire program. They had nothing but praise for our effort and progress, but I was extremely put out by their visit. Never before in the entire history of the Skunk Works had we been so closely supervised and directed by the customer. "Why in hell do we have to prove to a government team that we knew what we were doing?" I argued in vain to Jack Twigg, our assigned Air Force program manager. This was an insult to our cherished way of doing things. But all of us sensed that the old Skunk Works valued independence was doomed to become a nostalgic memory of yesteryear, like a dime cup of coffee.

We had lived and died by fourteen basic operating rules that Kelly had written forty years earlier, one night while half in the bag. They had worked for him and they worked for me:

1. The Skunk Works program manager must be delegated practically complete control of his program in all aspects. He should have the authority to make quick decisions regarding technical, financial, or operational matters.
2. Strong but *small* project offices must be provided both by the military and the industry.
3. The number of people having any connection with the project must be restricted in an almost vicious manner. Use a small number of good people.
4. Very simple drawing and drawing release system with great flexibility for making changes must be provided in order to make schedule recovery in the face of failures.

5. There must be a minimum number of reports required, but important work must be recorded thoroughly.

6. There must be a monthly cost review covering not only what has been spent and committed but also projected costs to the conclusion of the program. Don't have the books ninety days late and don't surprise the customer with sudden overruns.

7. The contractor must be delegated and must assume more than normal responsibility to get good vendor bids for subcontract on the project. Commercial bid procedures are often better than military ones.

8. The inspection system as currently used by the Skunk Works, which has been approved by both the Air Force and the Navy, meets the intent of existing military requirements and should be used on new projects. Push basic inspection responsibility back to the subcontractors and vendors. Don't duplicate so much inspection.

9. The contractor must be delegated the authority to test his final product in flight. He can and must test it in the initial stages.

10. The specifications applying to the hardware must be agreed to in advance of contracting.

11. Funding a program must be timely so that the contractor doesn't have to keep running to the bank to support government projects.

12. There must be absolute trust between the military project organization and the contractor with very close cooperation and liaison on a day-to-day basis. This cuts down misunderstanding and correspondence to an absolute minimum.

13. Access by outsiders to the project and its personnel must be strictly controlled.

14. Because only a few people will be used in engineering and most other areas, ways must be provided to reward good performance by pay not based on the number of personnel supervised.

Although most of our cherished rules were now in tatters, my guys managed to finish their work on Have Blue in mid-November, nearly three weeks before the flight test target date of December 1, 1977. "Rich," Bob Murphy teased, "you'd never have made your deadline by using regular workers. You had the cream of the crop in management delivering the goods for you." The airplane was loaded onto a C-5 cargo plane at two in the morning and roared away to our remote test site, leaving behind several complaints to the FAA from irate citizens whose sleep was disturbed by this violation of late-night takeoffs from the Burbank airport. Frankly, it was such a relief to get Have Blue out of assembly that I would have gladly paid a fine.

The plane was now in the hands of our flight test crews, who would spend the next couple of weeks performing flight control, engine, and taxi tests. Even though the test site was in a remote location, our airplane was kept under wraps inside its hangar most of the time. Soviet satellites made regular passes, and every time our airplane was rolled out everyone on the base who wasn't cleared for Have Blue had to go into the windowless mess hall and have a cup of coffee until we took off.

Seventy-two hours before the first test flight, the airplane began to seriously overheat near the tail during engine test runs. The engine was removed, and Bob Murphy and a helper decided to improvise by building a heat shield. They noticed a six-foot steel shop tool cabinet. "Steel is steel," Murphy said to his assistant. "We'll send Ben Rich the bill for a new cabinet." They began cutting up the cabinet to make the heat shield panels between Have Blue's surface and its engine. And it worked perfectly. Only in the Skunk Works . . .

It's the first of December, 1977, just after sunup, the best time for test pilots to take off. Winds are usually calmest then, but this morning the wind chill blasts through my topcoat like it's tissue paper. I'm wondering how I can be so damned cold while I'm sweating bullets over this test flight — probably the most critical test of my career. This flight will be every bit as important to the nation's future and the future of the Skunk Works as the first test flight of the U-2 spy plane, which took place at this very same highly secret sand pile more than a quarter century ago.

Back then, I was a Skunk Works rookie and this base, which we built for the CIA, was just a tiny outpost of windswept quonset huts and trailers, guarded by rookie CIA agents with tommy guns. Kelly had jokingly nicknamed this godforsaken place Paradise Ranch, hoping to lure young and innocent flight crews to work on a dry lake bed where quarter-inch rocks blew around most afternoons. It is now a sprawling facility, bigger than some municipal airports, a test range for sensitive aviation projects. No one nowadays gains access without special clearances that include a polygraph test. Such paranoia has kept our most guarded national defense secrets secret.

I've been here many times over the years on many Skunk Works test flights, usually accompanying Kelly Johnson. Today, the Have Blue prototype that will soon be rolling down this runway is the first built under my regime after Johnson's retirement three years earlier. But we really aren't one hundred percent certain that this sucker can actually get off the ground. It is the most unstable and weirdest-looking airplane since Northrop's Flying Wing, built on a whim back in the late 1940s.

I watch nervously as Have Blue emerges from the guarded cavity inside its hangar and is rolled out. It is a flying black wedge, carved out of flat, two-dimensional angles. Head on, with its black paint and highly swept wings, it looks like a giant Darth Vader — the first airplane that has not one rounded surface.

Bill Park, our chief test pilot, complained that it was the ugliest airplane he'd ever strapped himself into. Bill claimed that *flying such a mess earned him the right to double hazard pay. I agreed. He's getting a $25,000 bonus for this series of Have Blue test flights. To Bill, even the opaque triangular cockpit is ominous, especially if he has to punch out. But the specially coated glass will keep radar beams from picking up his helmeted head. The real beauty of Have Blue is that Bill's head is a hundred times more observable on radar than the airplane he will be flying.*

The sharp edges and extreme angular shape of our small prototype create whirling tornadoes and make the airplane a flying vortex generator. To be able to fly at all, the airplane's fly-by-wire system must operate perfectly, otherwise Have Blue will tumble out of control.

I check my watch. Nearly 0700. I give the thumbs-up sign to Bill Park in the cockpit, who's preoccupied with last-minute preflight checks. Kelly Johnson is standing at my side, looking stoic. He's still skeptical about whether or not this prototype will prove way too draggy to get off the ground. But Kelly brought along a case of champagne on the Jetstar from Burbank to celebrate after Park's flight. Over the years at the Skunk Works we've never failed to celebrate a successful maiden test flight of anything we've ever built. We always polished off a hard-earned success with a boisterous party where Kelly challenged all comers to an arm-wrestling contest. He's an old man now, ailing, but I still wouldn't take him on.

We've had our share of crashes during long weeks and months of test-flying new airplanes, but they didn't really upset us too much as long as no one got hurt, because we always learned important lessons from mistakes. But we never had a mishap on a first test flight — a catastrophe that would send us back to the drawing board with our tails between our legs.

Adding to the tensions of this day, the White House Situation Room is monitoring this flight. So is the Tactical Air Command at Langley Air Force Base in Virginia. But my anxieties are closer to home: I've got ten million bucks of Lockheed's money riding on this flight and the success of this program. I'm the one who

talked our board into going along with me. So I didn't need any
black coffee this morning. I am wired.

Bill Park fires the twin engines. The airplane has a muffled
sound because its engines are hidden behind special radar-
absorbing grids. Bill has been practicing using these flight con-
trols under all conditions in a simulator for five weeks and I
know he's ready for any emergency. He and I have been through
tight spots on other test programs. Once he ejected from an SR-71
that began to flip over on takeoff. I was sure Park was about to
become a grease spot on the tarmac, but his chute opened just as
his feet hit the ground, yanking him upward as he was impact-
ing. He left three-inch-deep heel imprints in the sand, but was
unhurt. Bill is damned thorough and damned lucky, a great com-
bination for someone in his line of work.

Kelly Johnson is watching intently as the prototype taxis past
us heading for the end of the runway, where it will turn and take
off into the stiff wind. Suddenly, in a blast of loud noise, the med-
evac chopper, with two paramedics on board, takes off and heads
down range to be in position if Park augers in. It is followed by a
T-38 jet trainer carrying one of Bill's test pilot colleagues, who
will fly chase, visually monitoring his airplane and supplying any
help or advice in an emergency.

Bill pushes on the throttle, and Have Blue slowly begins to
accelerate. To stay stealthy Have Blue has no afterburner, and it
will need almost as much runway to get airborne as a 727 loaded
with fuel, baggage, and passengers bound for Chicago.

Bill goes full throttle. He's chewing up a lot of runway as he
sweeps past us. I'm thinking, Damn it, with all that wind he
should be up by now. He's far down the runway and I'm no
longer breathing. Uh-oh. He's damn near off the end of the god-
dam runway. Then I see him lift off. Slow as a jumbo jet a hun-
dred times its weight, but he's up. His nose is high. But just
hanging there. Get up. Up, up, up. The little airplane hears me.
It's heading toward the snow-powdered mountains. Ken Perko of

the Pentagon's Advanced Research Projects Agency, who is among the half dozen outsiders cleared to witness this flight test, reaches out to shake my hand. "By God, Ben," he says, "the Skunk Works has done it again."

Kelly slaps me on the back and shouts, "Well, Ben, you got your first airplane."

Not so fast. It's standard procedure to leave the landing gear down on maiden test flights checking out airworthiness, but even so it seems to me the airplane is way too sluggish gaining altitude. There are some significant foothills looming in Bill's flight path and I try to do some quick mental calculating to get him safely over the hump. I raise my binoculars and quickly try to adjust the focus. By the time the mountains come clear, our airplane is across the other side.

Other Voices
Bill Park

Most people think of test-flying from old movies, where the girl and the pilot's best friend are watching the skies as he adjusts his goggles and starts the fatal dive. If the movie was a romance, the pilot usually made it. One way or another the flight test of a new airplane was over after one hair-raising dive.

It should only be that easy. We built two Have Blue prototypes in record time, only twenty months from the day the contract was awarded until I made the first flight. But the intensive flight testing of these two revolutionary airplanes took us two years. We needed a year or more to work out all the kinks — thoroughly evaluating the structural loads, performance characteristics, flight controls, avionics — and then make all the fixes. The next phase would be to test Have Blue against highly calibrated radar systems and precisely measure its stealthiness from every angle and altitude and be challenged by the most sophisticated radar systems in the world. That phase too

would take more than a year. Then the Air Force would evaluate the results and determine whether or not to go ahead with
full-scale production.

The Skunk Works gave its flight test group unique responsibilities: we had our own engineers, who had worked side by
side with fuel systems engineers, hydraulic specialists, the
landing gear team, as the airplane was being assembled. We
knew every nut and bolt long before first flight — a big edge
when the time finally came to push that throttle.

I was the principal pilot on Have Blue. My backup was a
blue-suiter, Lt. Colonel Ken Dyson. We didn't know very much
about the airplane in the beginning. It was built on the cheap
all the way. It was just a demonstrator that was to be junked, so
the brakes were god-awful, the cockpit too small and too
crammed. All the avionics were surplus store red tags. I remember this Air Force colonel came down to the test site and
asked me how much we spent on this program. I told him $34
million. He said, "No, I don't mean one airplane. I mean both
airplanes — the entire program." I repeated the figure. He
couldn't believe it.

The airplane was officially called the XST — the experimental stealth technology testbed. It was a dynamic laboratory in a
controlled environment. Everyone briefed on the program
knew full well the potential implications of this prototype for
the Air Force's future. If this airplane lived up to its billing, we
were making history. Air warfare and tactics would be changed
forever. Stealth would rule the skies. So everyone involved in
the testing was impatient to get test data, but it was my ass on
the line if something went wrong. And I wasn't about to risk it
by cutting any corners or rushing into test flights prematurely.

A helicopter with a paramedic on board was always airborne whenever I was doing test flights. And by May 1978, a
year and a half into the program, with about forty flights under
my belt, we were on the verge of graduating into the next phase

and beginning actual testing against radar systems. On the morning of May 4, 1978, Colonel Larry McClain, the base commander, stopped me at breakfast to say he would be flying chase for me that day and wanted to scrub the paramedic from the test flight because he needed him at the base clinic. I shook my head. I told him, "I'd rather you didn't do that, Colonel. We're not entirely out of the woods yet with Have Blue, and I'd just feel better knowing that paramedic is standing by if I happen to need him."

As it turned out, I had just saved my own life.

A couple of hours later I was completing a routine flight and coming in for a landing. I came in at 125 knots, but a little high. I was just about to flare and put the nose down when I immediately lost my angle of attack and the airplane plunged seven feet on one side, slamming onto the runway. I was afraid I'd skid off the runway and tear off the landing gear, so I decided to gun the engines and take off and go around again. I didn't know that that hard landing had bent my landing gear on the right side. When I took off again, I automatically raised my landing gear and came around to land. Then I lowered the gear, and Colonel McClain, my chase, came on the horn and told me that only the left gear was down.

I tried everything — all kinds of shakes, rattles, and rolls — to make the right gear come down. I had no way of knowing it was hopelessly bent. I even came in on one wheel, just kissed down on the left side, hoping that jarring effect would spring the other gear loose — a hell of a maneuver if I have to say so — but it proved useless.

By then I was starting to think serious thoughts. While I was climbing to about 10,000 feet, one of my engines quit. Out of fuel. I radioed, "I'm gonna bail out of here unless anyone has any better idea." Nobody did.

I would've preferred to go a little higher before punching out, but I knew I had to get out of there before the other engine flamed out too, because then I had all of two seconds before we'd spin out of control.

Ejecting makes a big noise — like you're right up against a speeding train. There was flame and smoke as I got propelled out. And then everything went black. I was knocked unconscious banging my head against the chair.

Colonel McClain saw me dangling lifelessly in the chute and radioed back, "Well, the fat's in the fire now." I was still out cold when I hit the desert floor face down. It was a windy day and I was dragged on my face by my chute about fifty feet in the sand and scrub. But the chopper was right there. The paramedic jumped out and got to me as I was turning blue. My mouth and nose were filled with sand and I was asphyxiating. Another minute or two and my wife would've been a widow.

I was flown to a hospital. When I came to, my wife and Ben Rich were standing over my bed. Ben had flown her in from Burbank on the company jet. I had been forced to bail out four times over fifteen years of flight testing for the Skunk Works, and I never suffered a scratch. This time I had an awful headache and a throbbing pain in my leg, which was in a cast. A broken leg was not fatal in the test flight business but my pounding headache was. I had suffered a moderate concussion and that was the end of the line for me. The rules were very strict about the consequences of head injuries to professional pilots. My test-flying days were over. Ben named me chief pilot, putting me in charge of administrating our corps of test pilots. Lt. Colonel Ken Dyson took over the Have Blue tests. He flew sixty-five sorties against the radar range with the one remaining prototype. On July 11, 1979, he got two hydraulic warning lights about thirty-five miles from base. Knowing he was flying a plane with no stability if the power went, he got out before it spun out of control. Ken parachuted safely to the desert floor. At the time of the crash, he had only one more scheduled flight and most of the test results were already in.

Have Blue flew against the most sophisticated radars on

earth, I think, and broke every record for low radar cross section. At one point we had flown right next to a big Boeing E-3 AWACS, with all its powerful electronics beaming full blast in all directions. Those guys liked to brag that they could actually find a needle in a haystack. Well, maybe needles were easier to find than airplanes.

3

THE SILVER BULLET

MY STYLE OF LEADERSHIP at the Skunk Works was markedly different from Kelly Johnson's, and it was wryly described by John Parangosky, the CIA's program manager for several Skunk Works projects, who knew us extremely well: "Kelly ruled by his bad temper. Ben Rich rules with those damned bad jokes." I was ebullient, energetic, a perennial schmoozer and cheerleader with an endless supply of one-liners and farmer's daughter jokes supplied fresh daily by my brother, a television producer on a situation comedy. Being so "user-friendly" was in sharp contrast to Kelly, who seldom made small talk and expected crisp, informed responses from his senior people to his sharp, pointed questions. When younger employees happened to see Kelly heading their way, they often dove for cover. I believed in the nonthreatening but benignly authoritarian approach to maintain high morale and team spirit. I spent half my time complimenting my troops and the other half bawling them out. Of course, by 1978, I was bouncing on pink clouds, enjoying the hosannas reserved either for angels or the head of a research and development outfit that produced a technology everyone wanted. Producing a new technology was the R & D equivalent of scaling Mount Everest. Northrop, our closest rival in developing stealth, was very good, but we were significantly better, and I was now taking meetings with admirals and four-star generals from all branches, each eager to buy into the new technology for their tanks and shells and missiles.

Rolling small ball bearings across the desks of four-star generals had paid off handsomely. "Here's the observability of your airplane on radar," I declared to their astonishment. By contrast, most fighters in the current inventory had the radar signature of a Greyhound bus, so the Air Force could not wait to shrink to marble size and signed a contract with us to start engineering a stealth fighter in November 1977, one month ahead of Have Blue's first flight test.

I was thunderstruck. We were rewarded with a development contract for a new fighter before our Have Blue demonstrator actually proved it could fly. No one in the defense business would be able to recall an occasion when the blue-suiters pulled an end run around their own inviolate rule: "Fly it before you buy it."

Military aircraft were so expensive and complex and represented such a sizable investment of taxpayers' money that no manufacturer expected to win a contract without first jumping through an endless series of procurement hoops, culminating in the flight-testing phase, that under normal circumstances stretched nearly ten or more years. From start to finish, a new airplane could take as long as twelve years before taking its place in the inventory and become operational on a flight line long after it was already obsolete. But that was how the bureaucracy did business. Within the Air Force itself, the decision to proceed on a particular project usually followed months, sometimes years, of internal analysis, debate, and infighting, which ensured that every new airplane was designated for a very specific operational purpose.

In our case the airplane was untested and its strategic purpose unclear. But William Perry, the Pentagon's chief for research and engineering, who had come into office with the new Carter administration in January 1977, took one look at the historic low observability results we achieved and immediately set up an office for counter-stealth research to investigate whether or not the Soviets had ongoing stealth projects;

the CIA began an intensive search to find out what the Russians were doing in stealth technology by redirecting satellites to overfly their radar ranges. The agency concluded that their only real interest in stealth was some preliminary experiments with long-range missiles. Otherwise, stealth was not a priority for them. Why spend money on a costly stealth delivery system when the U.S. had so few defensive missile systems and none nearly as sophisticated as their own?

The Soviets' apparent indifference to stealth spurred Bill Perry into action. In the spring of 1977, he called in General Alton Slay, head of the Air Force Systems Command. "Al," he said, "this stealth breakthrough is forcing me into a snap decision. We can't sit around and play the usual development games here. Let's start small with a few fighters and learn lessons applicable to building a stealth bomber."

The Air Force, like a shopper, bought by the pound: the lighter the cheaper. The rule of thumb was that the airplane's structure cost roughly a thousand dollars a pound, while its avionics were prime cut — four thousand dollars a pound at 1970s prices. Had Perry immediately pushed for a stealth bomber, General Slay would probably have done all in his considerable power to kill it. Not because he opposed stealth, but he was then up to his eyeballs trying to make Rockwell's troubled B-1A bomber live up to its advance billing as the successor to the B-52 long-range bomber. The B-1 was his number one priority. He very quickly got word sent to me via a subordinate: "Tell Ben Rich not to lobby around about a stealth bomber." He was one tough hombre.

In early June, Dr. Zbigniew Brzezinski, President Carter's NSC chief, whom I had never before met, decided to fly out to see Have Blue for himself. Brzezinski flew in an unmarked private jet to the remote base where I awaited him inside a tightly guarded, closed hangar. We spent several hours together. I let him kick Have Blue's tires and peer into the cockpit. Inside a secure conference room next to the hangar, I briefed Brzezinski on the stealth program and he began to question me: "How much stealth is enough stealth?" "Could stealth be applied to a conventional airplane without having to start from scratch?" "How long would it take the Russians to duplicate our stealth diamond shape if a model fell into their hands?" "How long before the Russians are likely to produce counter-stealth weapons and technology?"

Brzezinski scribbled my replies on a small pad. Then he asked me about the possibilities for developing a stealthy cruise missile that could be air-launched from a bomber and overfly unseen two thousand miles or more inside the Soviet Union to deliver a nuclear punch. I told him our preliminary design people were already at work on developing such a missile, which would be basically the same diamond shape as Have Blue. But without a cockpit in the configuration, the stealthiness was almost an order of magnitude better than even Have Blue — making our cruise missile design the stealthiest weapon system yet devised.

I showed him a copy of a threat analysis study prepared for us by the Hughes radar people, who were the best in the business, predicting near invulnerability for a stealthy cruise missile attacking the most highly defended Soviet target versus only a probable 40 percent survivability rate for the B-1A bomber. He asked for a copy of the study, a photo of the Have Blue airplane, and design drawings of the cruise missile to show to President Carter.

As he was leaving, Brzezinski asked me a bottom-line question: "If I were to accurately describe the significance of this stealth breakthrough to the president, what should I tell him?"

"Two things," I replied. "It changes the way that air wars will be fought from now on. And it cancels out all the tremendous investment the Russians have made in their defensive ground-to-air system. We can overfly them any time, at will."

"There is nothing in the Soviet system that can spot it in time to prevent a hit?"

"That is correct," I replied with confidence.

Three weeks later, on June 30, 1977, the Carter administration cancelled the B-1A bomber program. I had no doubt there was a direct cause-effect relationship between our stealth breakthrough and scrubbing the new conventional bomber. When I heard the news, I knew there would be at least one powerful Air Force general hopping mad and looking for someone to blame. I buzzed my secretary and told her, "If General Slay calls, tell him I'm out of the country."

It would be several months before I received any definitive word from the government on how they would proceed on stealth. During this long silence, I would later learn, a behind-the-scenes debate raged among the top echelons of the Air Force and the Defense Department on the best uses of stealth to provide us with the maximum strategic advantage against the Soviet Union. Within the Air Force the debate was between the Strategic Air Command, furious at losing its B-1 bomber, and the Tactical Air Command, eager to add a stealth fighter to its inventory. The referees in the middle were Secretary of the Air Force Hans Mark, an atomic physicist and former director of NASA's Ames laboratory, who was skeptical about stealth and a strong advocate of promoting missiles over manned bombers, and General David Jones, then chairman of the Joint Chiefs of Staff, who kept his powder dry and his opinions to himself until he was asked to make a decision. In the end it was General Jones who displayed the wisdom of Solomon: he gave SAC the green light to proceed with developing our cruise missile, and he approved the stealth fighter.

General Bob Dixon, head of the Tactical Air Command, flew out to see me in Burbank. "Ben," he said, "we want you to build us five silver bullets for starters. We'll take twenty more down the line."

In the jargon of the trade, a silver bullet was a deadly secret weapon kept under tight wraps until it was ready to be used to take out an enemy in a Delta Force covert surgical strike. The Israeli air force hit against Saddam Hussein's nuclear bomb facility in Baghdad was the perfect example of a Delta Force–style surgical strike operation. The silver bullet would be used to quick-hit the highest-priority, heavily defended targets in the dead of night.

Actually, it was an ideal Skunk Works project: tightly secret, building small numbers of hand-made airplanes rather quickly and efficiently. But I knew we also faced a steep learning curve leaping from building the small Have Blue demonstrator, with its off-the-shelf avionics, to a truly sophisticated larger fighter with novel and complex avionics and weapons systems.

Not long after General Dixon's visit, the chief of staff himself detoured from some business he had in San Diego, to drop by before going back to Washington. Among the services, the Navy was the most active in running "deep black" programs, especially in Navy SEAL penetrations of Soviet harbor and naval installations. But as General Jones reminded me over sandwiches in my office, "Your stealth fighter is the first black program the Air Force has ever run. Security is paramount. I doubt there are ten people in Washington aware of this project. Maintaining secrecy must be your number one priority, even ahead of keeping to the schedules and so forth. A leak in the papers would be disastrous. Be prepared to sacrifice efficiency or anything else to maintain the tight lid. Do that, Ben, and you'll keep out of trouble. The payoff for this airplane will be total surprise on the enemy the first time it is used." The president wanted Jones to personally brief Secretary of State Cyrus Vance and Defense Secretary Harold Brown on Have Blue and the other stealth projects. I had a briefing book prepared, which he took with him back to Washington. Before he left, the general told me that Admiral Bobby Inman, head of the supersecret National Security Agency, which operated all U.S. satellite and communications monitoring activities, was being brought into our stealth project to take direct charge of communications security between the Skunk Works, the test site, and the Pentagon. We would be receiving special cryptographic gear and scrambler fax and telephone systems.

I made a mental note that General Jones was not the one to complain to when Air Force security began driving me up a wall.

By my third anniversary since taking over from Kelly Johnson in 1975, the Skunk Works had added one thousand new workers and by 1981 would employ seventy-five hundred. Our drafting rooms and workshops were operating on overtime; our assembly hangars hummed around the clock, on three shifts. In addition to stealth, we were updating squadrons of older Blackbird spy planes, now twenty-five years old, with new wiring and avionics. We were also building six brand-new TR-1 spy planes a year, for a total of thirty-five, the deal I had closed with General Jones the first year of my regime. I was happily putting in twelve- to fourteen-hour workdays and so was nearly everyone else. Still, as a businessman I believed in the adage of "strike while you're still hammering" — and I pitched the Pentagon for seed money to develop stealthy helicopter rotor blades and anything else we could think up. Some wags in my employ presented me with a bowling ball stamped TOP SECRET. The attached card explained it was the equivalent of the radar cross section of the Pentagon, once we diamond-shaped it. The instructions said to roll it across the desk of the secretary of defense.

I should have been in high clover instead of up to my lower lip in deep doo-doo, but General Al Slay did get the last word and a measure of revenge for the loss of his beloved B-1: he forced on us a contract that was almost punitive. Because the Air Force had gone the unusual route of contracting for an airplane before the technology was proven in flight test, I was being socked with a contract worth $350 million to deliver the first five stealth fighters under draconian terms that could absolutely ruin us. Ultimately I had to guarantee that the stealth fighters would meet the identical radar cross section numbers achieved by our thirty-eight-foot wooden model at the White Sands radar range in 1975. I had also to guarantee performance, range, structural capability, bombing accuracy, and maneuverability.

The contract was like a health care insurance policy without catastrophic coverage: you were fine as long as you were fine. If something terrible happened, you would go down the tubes dead broke. If it proved impossible for us to duplicate the incredible invisibility of a wooden model with a full-size flying machine, we would be penalized and expected to foot the entire bill to get it right. I was feeling particularly skittish on that score because a few weeks before the contract negotiations began, I received an urgent call from Keith Beswick, head of our flight test operation out at the secret base.

"Ben," he exclaimed, "we've lost our stealth." He explained that Ken Dyson had flown that morning in Have Blue against the radar range and was lit like a goddam Christmas tree. "They saw him coming from fifty miles."

Actually, Keith and I both figured out what the problem was. Those stealth airplanes demanded absolutely smooth surfaces to remain invisible. That meant intensive preflight preparations in which special radar-absorbent materials were filled in around all the access panels and doors. This material came in sheets like linoleum and had to be perfectly cut to fit. About an hour after the first phone call, Keith phoned again. Problem solved. The heads of three screws were not quite tight and extended above the surface by less than an eighth of an inch. On radar they appeared as big as a barn door!

So the lesson was clear: building stealth would require a level of care and perfection unprecedented in aerospace. The pressure would really be on us to get it right the first time or literally pay a terrible price for our mistakes. Deep down, I felt confident that the Skunk Works would rise to the challenge. We always had in the past. Still, I had to swallow hard taking my case to our corporate leaders, who were still struggling to put our company back on its feet. They reacted with about as much apprehension as Kelly Johnson had when I told him about the contract: "Oh, boy. You could wind up losing your ass."

I argued that management had expected me to hustle and get new business, which also meant taking risks. Our new CEO, Roy Anderson, and president Larry Kitchen were clearly worried about our ability to duplicate the low radar cross section we had achieved on a small wooden model. "That is just asking for big league trouble promising to equal that," Kitchen remarked. I couldn't deny that he was right. I said, "We've already shown that we know what we are doing when it comes to stealth. We've been as good as our predictions up to now. And there's no reason to think we'll drop the ball. We'll build up a quick learning curve delivering these first five airplanes, and if we do hit a snag, we'll make it up off the back end. The fifteen to come will provide our profit margin."

One or two executives wanted me to refuse the deal and wait for the end of the Have Blue tests in the next year or so, when the Air Force would not be so intent on covering their own butts because they were buying untested merchandise. I rejected that idea. "Right now, we've got a contract and also the inside track on the next step, which is where the big payoff awaits: building them their stealth bomber. That's why this risk is worth taking. They'll want at least one hundred bombers, and we'll be looking at tens of billions in business. So what's this risk compared to what we can gain later on? Peanuts."

It was not a very happy meeting, and the conclusion reached was reluctant and not unanimous. The corporate bean counters insisted we install a fail-safe monitoring and review procedure that would sound the alarm the moment we fell behind or hit any snags. "Above all, no nasty surprises, Ben," Larry Kitchen warned me. Frankly, he sounded more prayerful than hopeful.

From that moment on, a hard knot formed in my gut around my biggest worry: guaranteeing bombing accuracy. Who knew what huge, ugly, time-consuming problems lay in store for us solving that one? Unlike the low-flying B-1 bomber that attacked from the deck, we would come in relatively high — twenty thousand feet or more — giving us a tighter circle to aim at. Also, because we would be invisible, our pilots would not have to duck and weave to avoid missiles or flak. We would have a clear shot to drop a pair of two-thousand-pounders. Hopefully, laser-guided smart bombs sighted by the pilot in the cockpit would prove unerring. Otherwise, I was in the tank until we found out how to make those damned bombs wise up.

The Air Force pressured me to accept a deadline of twenty-two months to test-fly the first fighter. It had taken us eighteen months to build Have Blue, which was far simpler, but I reluctantly agreed to meet the deadline. As Alan Brown, my program manager for the fighter production, put it, "Ben said 'Okay.' The rest of us said, 'Oh, shit.'"

The contract was signed on November 1, 1978. We had only until July 1980 to build the first airplane, get it right, and get it flying.

Kelly Johnson had operated under tremendous pressure on a lot of projects over the years, but he never had to put up with the galloping inflation that hit us unexpectedly in 1979 as the OPEC oil cartel suddenly raised prices more than 50 percent. Sixteen percent inflation rates were eating me alive, and my contract with the Air Force had no price-adjustment clauses to relieve some of the financial pressures. "Who could've foreseen this goddam mess?" I howled to the winds. Our accounting office was becoming apoplectic. The Air Force sympathized and told me to keep my chin up but rejected my appeal for renegotiations to build inflationary spirals into a shared customer-government cost outlay. By the middle of the presidential campaign of 1980, Carter was catching hell from all directions. Ronald Reagan blasted him for weakening the military and made a campaign issue out of Carter's cancellation of Rockwell's B-1 bomber, which had cost eight thousand jobs in voter-rich Southern California. The Carter White House asked me to draft a briefing paper for Reagan that would privately inform him about the very sensitive stealth project in the hope he would back off his attacks on the outmoded B-1. Fat chance that would happen, but in a desperate move, Defense Secretary Brown shocked me by stating in public that the government was doing research on important stealth technology. By then Carter had lost the defense issue totally, so Brown should have kept his mouth shut.

We in the Skunk Works had done very well under the Carter administration and would really miss tremendous performers like Bill Perry at the Pentagon.* But Reagan roared into Palmdale and blistered Carter with a speech at the Rockwell plant, promising to reopen the B-1 bomber line after the election. Everyone in aerospace was ready for a change. Guys in the plant were whistling "Happy Days Are Here Again" simply because the sentiment fit perfectly with their mood. The so-called Misery Index, cited by Reagan, which was the rate of inflation measured against declining employment, really resonated with me. I felt that Misery Index every time I sat down with our auditors and watched my costs slam through the roof.

In one of his final acts before leaving office, Defense Secretary Harold Brown called me to Washington on the eve of

---

* Perry is now serving in the Clinton administration as the Secretary of Defense.

Reagan's inauguration in January 1981, and in a secret cere-
mony in his Pentagon office awarded me the Defense Department's Distinguished Service Medal for the stealth airplane.
Because of the tight security surrounding the project, only
Kelly was allowed to accompany me. He stood by beaming
like a proud uncle as Brown pinned on my medal and said,
"Ben, your Skunk Works is a national treasure. The nation is
in your debt for stealth and all the other miracles you people
have managed to pull off over the years. From all of us in this
building, thank you."

I was allowed to show the medal to my two children, Karen
and Michael, but I couldn't tell them why I had received it.

Reagan would initiate the biggest peacetime military
spending in our history. During the early 1980s defense industry sales increased 60 percent in real terms and the aerospace
workforce expanded 15 percent in only three years — from
1983 to 1986. We employed directly nearly a quarter million
workers in skilled, high-paying jobs and probably twice that
many in support and supplier industries. Not since Vietnam
were we building so much new military equipment, and that
fevered activity was, coincidentally, being matched in the
civilian airline industry.

Boeing, in Seattle, was reaping the biggest bonanza in its
history during the first years of the 1980s, filling orders from
the major airlines to invest in the next generation of 727s,
737s, and 747s. One airliner a day was rolling out of the huge
Boeing complex. Between Boeing and the growing production
lines for new missiles and fighters at California-based aerospace outfits, I suddenly found myself on the short end of materials, subcontracting work, machine shop help, and skilled
labor. Without warning, there was a dire shortage of everything used in an airplane. Lead times for basic materials
stretched from weeks to literally years.

We needed specialized machining and forgings, and our local subcontractors just shrugged us off. We were small

potatoes, who bought in threes and fours. We advertised our
needs as far away as Texas, usually in vain. Even a favorite
landing gear manufacturer for past projects had to turn us
down; he had no time to start up a production line for such a
small order. I even had to beg for aluminum — Boeing's huge
airliners were hogging the 30 percent of aluminum production allocated for the airplane industry. The remainder was
allocated to the soft drink and beer industry. I had to personally plead with the head of one of the Alcoa plants whom I
knew to stop a run and squeeze in our modest order. He did
me a personal favor — things were that tight.

Finding qualified aerospace workers was almost impossible
at any price. Usually we borrowed people from the main
plant, but business was brisk there, too, building our own Tri-
Star airliner and completing a big contract award for a Navy
patrol aircraft, and they had no skilled workers to spare. We
had to hire people off the street, and security clearances became a horror and a half. We'd find someone with good references as a welder only to have him flunk security because of
drugs. Forty-four percent of the people who applied for jobs
with us flunked the drug testing. I began to think that all
of Southern California was zonked on coke, heroin, pot,
and LSD. Those who flunked were mostly shop personnel,
but some promising technical types were caught in the net
as well.

We weren't exactly home free with many of the new employees who did pass the drug hurdle; we had to start from
scratch getting them cleared and it could take longer than
having a baby. I got dispensations from security for workers
we purposely put in "ice boxes" — that is, they worked in remote buildings far from the main action, assembling innoc-
uous parts. We were purposely creating big problems in terms
of efficiency and logistics in the name of security by allowing
ourselves to become so fragmented. But I had no choice. I had
to tuck away workers so they couldn't see or guess what it was

they were really working on. I had to make us inefficient by
having them work on pieces of the airplane that would not
reveal the nature of the airplane itself. I couldn't tell them how
many pieces they had to make, and we had to redo drawings
to eliminate the airplane's serial numbers. That alone required
significant extraneous paperwork. The majority of the people
we hired had no idea that we were building a fighter, or
whether we were building ten or fifty. Through a complex procedure we reserialized their piecemeal work when it came
into the main assembly.

I had to laugh thinking how Kelly would have reacted not
only to the security headaches but to the exasperating management regulations that never existed in his day. I might be
cleared for top secret, but I was also on a government contract
and that meant conforming to all sorts of mandatory guide-
lines and stiff regulations. Kelly had operated in a paradise of
innocence, long before EPA, OSHA, EEOC, or affirmative action and minority hiring policies became the laws of our land.
I was forced by law to buy two percent of my materials from
minority or disadvantaged businesses, but many of them
couldn't meet my security requirements. I also had to address
EEOC requirements on equal employment opportunity and
comply with other laws that required hiring a certain number
of the disabled. Burbank was in a high-Latino community and
I was challenged as to why I didn't employ any Latino engineers. "Because they didn't go to engineering school" was my
only reply. If I didn't comply I could lose my contract, its high
priority notwithstanding. And it did no good to argue that I
needed highly skilled people to do very specialized work, re-
gardless of race, creed, or color. I tried to get a waiver on our
stealth production, but it was almost impossible.

We had barely any experience working with new exotic materials being used for the airplane's outer skin. The radar-
absorbing ferrite sheeting and paints required special precau-
tions for the workers. OSHA demanded sixty-five different

masks and dozens of types of work shoes on stealth alone. I was told by OSHA that no worker with a beard was allowed to use a mask while spray coating. Imagine if I told a union rep that the Skunk Works would not hire bearded employees — they'd have hung me in effigy.

The Skunk Works facilities were old, many of them dating back to World War II, and even a myopic OSHA inspector would have had a field day finding inadequate ventilation or potentially unsafe asbestos insulation still in the walls. Our work areas were very skunky, ladders all over the place, lots of wiring to trip over, an oil slick or two. We had worked fast and loose from day one — with seldom an accident or a screwup. That was part of our charm, I thought. We were great innovators, rule benders, chance takers, and when appropriate, corner cutters. We did things like fuel airplanes inside an assembly area — a strictly forbidden act that risked fires or worse — to solve the problem of not having to move a very secret airplane into daylight to see if its fuel system leaked. Our people knew what they were doing, worked skillfully under intense pressure, and skirted hazards mostly by sheer expertise and experience. But as we grew, the skill level decreased and sloppiness suddenly became a serious problem.

Midway into the stealth fighter project we began experiencing foreign object damage (FOD) caused by careless workmen. This particular problem is familiar to all manufacturers of airplanes but had been practically nonexistent in our shop. Parts left inside an engine can destroy it or cost lives in fatal crashes. We've all heard about surgeons leaving sponges or clamps inside bodies — but I know of a case in the main Lockheed plant where a workman left a vacuum cleaner inside the fuel tank of an Electra. The vacuum cleaner began banging around inside the fuel tank at ten thousand feet and the pilot landed safely before disaster struck. A big problem with jets is keeping runways clear of debris that could be sucked into an engine. Break off an engine blade and it rips through an engine causing catastrophic damage. In our case, workers would crawl into a space with pens in their pocket, oblivious when one dropped out, or they would carelessly leave a bolt or screw inside an engine. One loose bolt left inside could cause us to replace an entire $2.5 million jet engine. Carelessness was costing us about $250,000 annually in repairs. We solved part of the problem by designing pocketless coveralls and installing a very strict parts and tool auditing system on the assembly floor. Our people had to account for every rivet and screw.

We also learned to keep a sharp eye to ensure that workers didn't try to save time or cut corners by using tools not designed for particular parts. Another concern: workers would screw up and damage a part, but instead of reporting it to their supervisor, they'd sneak off to the supply cabinet and grab another part that was reserved for the next plane they would be building. We learned to keep our parts locked and tagged so that workers could not obtain easy access. We also discovered that some of our welders and riveters had bypassed their required semiannual certification tests. The Air Force auditors were hound dogs and our record keeping stank. After decades of successfully avoiding red tape we were now swimming in it.

"Face it," I told my supervisors, "our people are getting too damn lax." We were working three shifts, around the clock, building the stealth fighter. When you build one or two airplanes at a time there isn't as much discipline as when you are building dozens. Our people never cleaned up their work areas before the next shift came on until I ordered them to stop working fifteen minutes before the next shift and use that time to sweep up and pick up.

The bottom line was that I was forced to use too many inexperienced workers. On the one side I had General Dixon of the Tactical Air Command climbing all over me because of foreign object damage and insisting that he bring in a team of efficiency experts to clean up the mess. "Ben, I know you hate me for it now," Dixon said, "but you'll thank me for it later." He was right on both counts. Ultimately our shops became spotless and models of their kind. But it took a lot of stress getting us there. On the other side I was fighting off OSHA inspectors clamoring to get inside the Skunk Works and possibly close down our operation.

A few workers complained because they heard that the new radar-absorbing materials were made out of highly toxic composites and became concerned for their health. The truth was we were very careful how we used hazardous materials, but because of proprietary considerations I could not reveal in public the composition of our materials, which our competitors would be as eager to discover as the Kremlin. In desperation I called the Secretary of the Air Force to get those OSHA inspectors off my back. I was told, that's too hot for us to tackle, thank you very much. So I called OSHA and told them to send me the same inspector who worked the Atomic Energy Commission — a guy cleared for the highest security and used to working with highly sensitive materials. This inspector came out and nickel-and-dimed me into a total of two million bucks in fines for no fewer than seven thousand OSHA violations. He socked it to me for doors blocked, improper ventilation, no backup emergency lighting in a workspace, no OSHA warning label on a bottle of commercial alcohol. That latter violation cost me three grand. I felt half a victim, half a slumlord.

But then an even more serious problem hit us. A disgruntled employee, bypassed for promotion, contacted a staff member on the House Government Operations subcommittee and accused the Skunk Works of lax security and claimed that we lost secret documents. His accusations were perfectly timed because an airplane model manufacturer named Testors was making a fortune with a model they called the F-19, claiming it was America's supersecret stealth fighter. They took the front end of our Blackbird, put a couple of engines on

it, and advertised it as the stealth fighter. They sold 700,000 of
these bogus stealths and Congress was livid. They wanted to
know how could we allow the government's most secret ongoing project to become a best-selling Christmas present. A couple of congressional committees wanted to send for me and
sock it to me in executive session, but the Air Force refused to
allow my appearance under any circumstances, citing extreme national security concerns. So Congress reached into
our board room, and Larry Kitchen was sent to the Hill as the
sacrificial lamb instead; he was browbeaten unmercifully before the House Subcommittee on Procedures and Practices.
Then the subcommittee's chairman, John Dingell, a feisty
Michigan Democrat, sent a few of his committee sleuths to
Burbank to investigate our security procedures. They ordered
an audit of all our classified documents from year one — and I
almost had a stroke. The first thing I did was drive over to
Kelly Johnson's house and grab back cartons of documents
and blueprints and God knows what else, all stored in Kelly's
garage. Kelly operated by his own rules. He said, "Damn it, if
they can't trust Kelly Johnson by now, they can go straight to
hell." For years Kelly made his own security rules, but now the
rules had changed drastically and were vigorously enforced
and unbending. I was sweating that we'd all wind up making
license plates at Leavenworth.

    Government auditors discovered some classified docu-
ments missing. The documents in question had been properly
shredded, but our logging was antiquated and no one recorded the date of the document destruction. It was a bureaucratic foul-up rather than any serious security breach, but tell
that to Congress. The government cut my progress payments
on the stealth fighter project by 30 percent until I could prove
to their satisfaction that I had taken specific steps to eliminate
security logging laxness and lost documents. From then on,
we were monitored unceasingly. Toward the end of the stealth
project I had nearly forty auditors living with me inside our

plant, watching every move we made on all security and contract matters. The chief auditor came to me during a plant
visit and said, "Mr. Rich, let's get something straight: I don't
give a damn if you turn out scrap. It's far more important that
you turn out the forms we require."

Those guys swarmed over us like bees on clover, checking
up on our payment schedules, investigating whether we
bought the lowest-priced materials and equipment from subcontractors, whether we really negotiated cost, tracked it,
worked hard to get the best deal for Uncle Sam with our suppliers. I had to double my administrative staff to keep up with
all these audits. For better or worse, we were stuck inside a
Kafkaesque bureaucracy demanding accountability for every
nut, screw, and bolt.

In between all these distractions and disruptions we were
trying to build an airplane. We started assembly the same
time as McDonnell Douglas started the F-18 fighter. They took
ten years to produce their first operational squadron of twenty
airplanes. We took only five years. And theirs was a conventional airplane, while ours was entirely revolutionary technology.

We began by refining our shape on the computer and then
constructing a full-scale wooden mock-up so that the exact
shape and fit of each critical facet panel and component could
be evaluated and any problems associated with new details
like the bomb bay could be identified and solved. We knew
that this slightly newer and larger shape would be as unstable
as the Have Blue aircraft — but would there be differences?
To find out, one of our aerodynamicists built a giant slingshot
that looked like a rock-hurling catapult right out of an old
Robin Hood movie, set it up on the third-floor ramp of a huge
assembly building the length of two and a half football
fields — and then fired off models of our new stealth shape
and took slow-motion film of how they fell to the ground, receiving a painless preview of what would happen if the real

airplane spun out of control. Security forced us to do this indoors rather than off a rooftop — but it worked perfectly.

Other Voices
Alan Brown

I was Ben's program manager. Building the stealth fighter, we had to tightrope walk between extreme care and Swiss-watch perfection to match the low radar observability claims of our original computerized shape. We didn't have the time, money, or personnel to build a flying Mercedes. But we couldn't allow even the tiniest imperfection in the fit of the landing gear door, for example, that could triple the airplane's radar cross section if it wasn't precisely flush with the body. So we took extra steps to hold in those doors and put on an extra coating of radar-absorbing materials.

We were well aware that what we were doing was outside the scope of normal engineering experience. We were dealing with radar cross sections lower by *thousands* not *hundreds* of orders of magnitude.

Many of the airplane's details required breakthrough engineering, particularly in the engine intakes and engine exhaust system. The exhaust especially gave us fits. It was complex, using baffles and quartz tiles to resist telltale heat signatures. To keep us as stealthy as possible, we used only infrared systems to get us to the target and aim our bombs. These systems emitted no electromagnetic signals but were vulnerable in stormy weather because water absorbs infrared energy. We gave up 20 percent in aerodynamic performance because of the flat plate design, which meant we would have to refuel in flight more often to get to our target and back. The F-117's range was twelve hundred miles.

I had anticipated propulsion problems, which we didn't have, but two of our biggest problems were how to keep the tailpipe from cracking and the data measurement systems from icing. The tailpipe set us back months. The problem was that a flat tailpipe, which we had to use, was not structurally sound under high pressure and easily cracked. We just couldn't find a solution and finally got General Electric's engine division to deal with it; they were expert in high temperatures and we adopted their design. The air data measurement system, called pitot probes, could have sunk the entire project if we couldn't perfect it. Doing so took us the entire two and a half years. These probes, which extended out the nose in stiletto shapes, recorded for the onboard computer static pressure, dynamic pressure, airspeed, angle of attack, and angle of sideslip so that the computer could make its microsecond flight adjustments. If those pitot probes iced up, the airplane would go out of control in two seconds flat. So ours had to be foolproof and, while jutting out from the airplane's nose, stealthy as well. How to heat these probes to keep them from icing without having them become conductive and act like antennas to radar or infrared devices was a problem that ate us alive. We finally developed a nonconductive heating wire the thickness of a human hair.

Another big problem was canopy glass. The pilot must be able to see out with no radar energy seeing in. The pilot's head would be hundreds of times larger on radar than his airplane. We had to develop coating materials that would pass out one without allowing in the other.

Occasionally we ran up against a problem that just didn't make any sense. For example, suddenly a special ferrite paint we used to coat the fighter's leading edges lost its radar-absorbing potency. We couldn't figure out what went wrong until one of our people decided to confer with DuPont, our supplier, and discovered that they had changed the way they made the paint without informing us.

Ben kept a close eye on all our problems, but he was never a second-guesser. The most striking thing about his leadership — especially in comparison to Kelly Johnson, who was totally hands-on with technical people — was that Ben let us do our jobs with a minimum of interference. His style wasn't to redesign our design of our engine the way that Kelly absolutely would have done, but to let us do our thing and smooth our way with the Air Force and Lockheed management. Yet the F-117A tactical fighter was every inch Ben Rich's airplane. If he hadn't pushed for it right from the outset, we would never have got into the stealth competition. He was the perfect manager — he was there for tough calls and emergencies. He would defend and protect us if we screwed up and keep us viable by getting new projects and more money from the Congress, convincing them and senior government officials about the value of stealth. He had a hunch and a vision — and it paid off handsomely.

By the summer of 1980, we were supposed to have flown the first of the five test airplanes but found ourselves way behind schedule. Too many unsolved problems kept my bean counters frazzled and worried. The first airplane's serial number was 780 — July 1980 — the date of our scheduled test flight that now seemed far over the horizon.

But I took heart from the fact that our learning curve improved almost daily, that we were solving technical problems that would make future stealth projects far easier to manage. But between the Air Force brass pressuring me on one side and the concerns expressed by Lockheed management on the other, the pressures were almost at the critical mass before a blowout.

Missing that July 1980 deadline for the first flight test of the F-117 wasn't the end of the world, but it made me apprehensive because I could not honestly report to anyone that the worst delays and problems were all behind us. Each day brought a fresh challenge or crisis, and I was doing a lot of tossing and turning instead of sleeping.

That summer of 1980 was for me the low point of my life,

professionally and personally. I was working myself into a frazzle, juggling projects and problems like some lunatic circus acrobat. My meetings began not long after sunrise and my workday ended well after dark. Some days brought great news about solving a particularly tough problem. Other days, the airplane project seemed hopelessly mired in a swarm of complications. The problem-solving line forming outside my office door grew longer by the day. And I had good people who didn't come to me for help unless they felt they had no other choice.

My wife, Faye, married to a workaholic for more than thirty years, was used to my late hours. But one night in early June she greeted me at the door looking pale and shaken, and all my problems and pressures at the Skunk Works became insignificant. She had just turned fifty and had gone in for a routine medical checkup. An ominous spot was discovered on her right lung. Faye had a long history of asthma, so bad at times that we kept a small oxygen tank at home, and I prayed that somehow that spot had something to do with her chronic asthma. No such luck. Faye was biopsied and immediately operated on for cancer. Her lung was removed. The doctor told me that he was sure he got all the cancer and that she should recover completely. She came home on August 1, and I took a week off to nurse her. Her recovery seemed slow but steady.

On Monday, August 18, I got home early. We had dinner. Afterward, we watched the news on television and Faye complained of weakness. I decided to call her doctor, but before I could get to the phone, she began struggling to breathe and started turning blue. I ran to get the oxygen. Then I gave her an injection of adrenalin, which we had kept on hand for her severe asthma attacks. She failed to respond and I ran to the phone and dialed 911.

The paramedics arrived in only minutes, but they were too late. Faye died in my arms from a massive heart attack.

I've blotted out the next days and weeks. I vaguely remember sobbing with my married son and daughter and receiving an emotional hug at the cemetery from Kelly Johnson, whose own wife, MaryEllen, was desperately ill from diabetes. MaryEllen and Faye were close friends, and MaryEllen was devastated by Faye's passing.

I decided my only hope for keeping sane was to plunge immediately into my work. My younger brother, who had recently divorced, moved in with me. And on the morning I returned to work I found a piece of paper on my desk. It was from Alan Brown, who was managing the program, and written on it was the date of my next birthday — June 18, 1981. "What's this?" I asked. "That's the date we test-fly the airplane," Alan replied. "The date is firm. In granite. Count on it." I gave him a wan smile, because right then the tailpipe problem was still throwing us for a loop and flight testing seemed over the hills and far away.

But on Thursday morning, June 18, 1981, our first production-model stealth fighter took off from our base on its maiden test flight. She flew like a dream.

Postscript on a Big Hit

The success of the stealth fighter did more than just bail me out. I had emerged unscathed even though we lost slightly more than $6 million on the first five production models. But the Air Staff was so pleased with the airplane that they decided to go for twenty-nine, then fifty-nine. I almost had them convinced to go for eighty-nine. After the first two batches of deliveries we achieved phenomenal efficiency. So much so that we made about $80 million on the deal. At one point I offered to give the government some of its money back because even in the Reagan years I was scared of being accused of making excessive profits. That was a federal offense, punishable with heavy fines. The Air Force told me it had no bookkeeping methods for taking back money, so I gave them $30 million worth of free engineering improvements on the airplane. We were able to make so much because we had perfected every aspect of our manufacturing techniques.

Stealth was our great good fortune and our earnings skyrocketed. The stealth fighter brought in more than $6 billion. Refurbishing the U-2 and the Blackbird brought in $100 million. By my fifth year I was heading a small, secret R & D outfit whose annual earnings placed it among the Fortune 500. Not bad. Not bad at all.

4

SWATTING AT MOSQUITOES

T HE MAJOR'S name was Al Whitley. He was a top F-100 fighter pilot from the Tactical Air Command and only months away from being promoted to lieutenant colonel. He had about a thousand hours of flying logged in, including combat in Vietnam, and was the first blue-suiter recruited for the new, secret squadron of stealth tactical fighters. Whitley arrived at the Skunk Works in February 1982, accompanied by two crew chiefs, to watch us building his airplane — our first production model. The official Air Force designation for the airplane was the F-117A. Like everything else concerning the stealth fighter, even its designation was classified.

By the time the airplane rolled off the line three months later, Al and his crew would know every wire, gauge, and bolt. They would be followed by all the other pilots and crew in that first squadron, who enjoyed the unique opportunity of actually being in on the production of the airplane they would soon be responsible for flying safely and effectively. Our purpose was to help them overcome fears of the unknown and achieve a level of confidence bred of expert knowledge of what their new airplane was all about. No other aerospace manufacturer came close to establishing such an intimate working relationship between builder and user.

Major Whitley had been selected by "Burner" Bob Jackson, a two-fisted Tactical Air Command colonel, who rounded up the most mature and experienced fighter jocks on active duty and gave each of them a two-minute briefing on what they might be doing if they said yes. All he told them was that they would be able to fly their butts off. There would be considerable family separation in the process and the work would be extremely classified. They had exactly five minutes to make up their minds.

Whitley needed only ten seconds. Now he sat in my office impatient to get his first look at a stealth fighter. I told him, "Keep in mind that to achieve stealthiness we had to commit a planeload of aerodynamic sins. What we came up with suffers just about every kind of unstable flight dynamics." Then I escorted him and his two crew chiefs onto the production floor to see the airplane for the first time. I watched those three Air Force guys exchange anxious looks, like just before a first attempt at the high diving board. "Boy, it sure is an angular son of a bitch, isn't it?" Whitley muttered, seeing that top secret diamond shape for the first time.

I smiled reassuringly. "Major," I said, "I guarantee you that by the time you are ready to strap in that cockpit, you'll enjoy one of the sweetest rides of your life." And I wasn't just blowing smoke. We were determined to make the F-117A the most responsive and pilot-friendly airplane in the inventory. My feeling was that any airplane that looked so alien had better be easy to handle.

We had already built five. But because the Air Force was in such a rush to form a squadron, the F-117A was very much a work in progress, forcing us to leapfrog the prototype testing phase, which was only then getting off the ground. We used these first five airplanes as guinea pigs to test aerodynamics and propulsion, knowing that changes would come with experience. We kept detailed records of every part in every stealth fighter so that when we made fixes we could facilitate these changes on the earlier airplanes.

Our technicians would work on flight lines and in the hangars for as long as the airplane remained in the inventory, solving problems for the Air Force mechanics. The airplane's special need to have absolutely smooth surfaces in order to maintain maximum stealthiness caused unusual stress for ground crews. After each flight the radar-absorbing materials had to be removed to gain entry to doors and service panels, then had to be meticulously replaced in time for the next mission. If the crew screwed up, they'd lose a plane and a pilot, because one neglected indentation exposed to enemy radar acted like a neon pointer. The process was called "buttering," using a special radar-absorbing putty we developed to coat uneven surfaces.

The Air Force initially ordered twenty-nine fighters. We built the first one in May 1981 and airlifted it out to our base for flight tests. The first flight confirmed a nagging doubt I had that we had made the twin V-shaped tails too damned small. Midway through that test program, one of the tails fluttered off. The test pilot was able to land after flying for several minutes while actually unaware of what happened. "I thought the airplane acted a little sloppy," he told me later. His chase plane pilot had warned him, "Hey, I see one of your tails in free fall."

We had to redesign the tail, which turned out to be 15 percent too small and too flexible for directional stability and control. Otherwise the airplane handled well.

Looking back, I am frankly amazed we didn't have many more major problems to fix than that one because, in truth, we were operating under chaotic conditions. Not only did we suffer all kinds of inefficiencies because of the tight security regulations, but most of the thousand production and shop workers building this airplane were starting from scratch at the Skunk Works. The best tribute to our homegrown training program was the astounding learning curve we achieved in the first couple of years. Building only two airplanes every three months, we enjoyed a better learning curve — 78 percent — than other manufacturers had reported while building twenty-five airplanes a month. The rule of thumb in the aerospace business was the more you build, the better you get at it. Our view was that efficiency was mostly the result of quality training, careful inspection, supervision, and high worker motivation. And we achieved these efficiencies in the face of a glaring shortage of trained workers as the Reagan defense spending program began to accelerate in the 1981–1984 time period. The shortage became so severe that we borrowed tooling people from as far away as Lockheed's Georgia plant; by 1985, our workforce totaled a record seventy-five hundred workers on a variety of stealth and nonstealth secret projects. During this same period, the aerospace industry in Southern California, including Hughes, Rockwell, McDonnell Douglas, Northrop, and Lockheed, had added about forty-five thousand workers to its payroll as military aircraft revenues peaked at $33 billion in sales by 1986. The era of big defense-related profits was at hand.

But there was always a price to pay when too many inexperienced workers were doing vital work on an airplane. On April 20, 1982, Major Whitley's stealth fighter was ready to take its Air Force acceptance flight out at the secret base. Whitley himself wanted to take the flight, but that was strictly against our rules. Our veteran test pilot Bob Riedenauer got the assignment. The airplane had performed perfectly during predelivery testing, but the night before the test flight we relocated a servomechanism from one equipment bay to another and rewired it. Riedenauer had barely lifted off the runway when he found to his horror that the wiring had accidentally reversed his crucial pitch and yaw controls. The airplane was only thirty feet off the deck when he flipped over backward and crashed on the side of the lake bed in a billowing cloud of dust. Bob was trapped in the cockpit and had to be cut free, sustaining serious leg injuries that kept him hospitalized seven long and painful months.

"Holy shit," Major Whitley exclaimed, "that could have been me." We were both extremely shaken, but I was also hopping mad. That nearly fatal mistake should have been caught in the inspection process. Clearly, such an oversight compounded the original rewiring error. The Air Force convened an accident review board and noted that we had instituted new safeguards and inspection procedures within forty-eight hours of the accident. But the Air Force remained confident of the product, and Major Whitley finally took off for the first time in October 1982, flying the second production model. In honor of his maiden voyage, I presented him with a cryptically worded plaque that had to get by our security censors: "In recognition of a significant event, October 15, 1982." Al laughed, but it would be six long years before he could finally explain to his wife and kids what in hell that plaque's inscription really meant.

"You kept your promise," Whitley said to me. "I had a slight anxiety attack rolling down that runway, but as soon as I was airborne and those wheels were sucked up, the ride was pure exhilaration."

The stealth fighter became operational one year later. By then, the Air Force had decided to expand its deployment on a global scale, for a total of fifty-nine stealth fighters, to comprise three squadrons of a special and secret wing. One squadron would be deployed to England, for coverage of Western Europe, the Middle East, the Soviet Union, and Eastern Europe. A second squadron would be sent to South Korea, to provide standby attack capabilities throughout Asia. The third squadron would be in training stateside. We delivered thirty-three airplanes by 1986 and the remaining twenty-six by the middle of 1990. We built eight a year at a fly-away cost of $43 million each. Stealth did not come cheap, but considering the revolutionary nature of the product and the enormous strategic advantages it afforded, the F-117A was the most cost-effective new weapons system in the inventory. The first stealth fighter squadron, composed of eighteen airplanes and a few spares, was ready to go to battle only five years after the initial Air Force go-ahead, suffering only one, nonfatal, crash in the process.

Other Voices
Colonel Alton Whitley

The F-117A was the nation's best-kept secret. Only a very small number of Air Force brass even knew that we existed. The Pentagon located us in one of the most desolate spots in North America, on a remote high-desert airstrip originally used by the Sandia National Laboratories for nuclear warhead testing. It was part of the Nellis Air Force Base test range, about 140 miles from Las Vegas, an uninhabited area of undulating plains and scrub with looming High Sierra foothills in the far distance. The nearest town, about twenty miles away, was called Tonopah. Only the Lord knows how many other secret government projects were tucked away in remote corners of that huge test range, the size of Switzerland, but we figured we were far from alone out there. Wild mustangs roamed freely through the desert scrub and galloped across our runways. Big scorpions scuttled around the dayrooms and inside the new hangars we built to hide our airplanes from Soviet satellites. Colonel "Burner Bob" Jackson saw an advertisement in the *Wall Street Journal* by Chevron, out of Canada, offering to sell its temporary cold weather trailer units from a discontinued oil patch for $10 million. Jackson flew up there and bought the whole thing for a million bucks and had it shipped down to Tonopah. And that became our first temporary housing. But over the years the Air Force poured $300 million into the base, building three runways, and transformed it into a very major facility, complete with gym and indoor pool.

Before that base was ready and before we had enough fighters ready to fly, our newly formed squadron took over a remote corner of Nellis Air Force Base and spent our time flying A-7 attack fighters. The A-7s became our cover. In early 1984, we deployed in A-7s to Kunsan Air Base in South Korea, to test our deployment procedures to the Far East ahead of the F-117A squadron that would be sent there. The word was purposely leaked that our A-7 fighters were carrying supersecret atomic antiradar devices that would render the airplane invisible to enemy defenses. To maintain the deception we outfitted each plane with old napalm canisters painted black and flashing a red danger light in the rear. It carried a radiation warning tag over an ominous-looking slot on which was printed: "Reactor Cooling Fill Port." When we deployed carrying these bogus devices, Air Police closed down the base and ringed the field with machine gun–toting jeeps. They forced all the runway crews to turn their backs on our airplanes as they taxied past and actually had them spread-eagled on the deck with their eyes closed until our squadron took off. Real crazy stuff. But the deception actually worked.

When we finally moved into Tonopah in 1984, we kept A-7s parked on the ramp so that Soviet satellites would think we were an A-7 base. But if their photo analysis experts were really on the ball, they would have picked up the double fencing around the perimeter, the powerful searchlights, television cameras, and sensing devices, all signs of unusually tight security. And I think they may have picked this up because satellite overpasses increased to as many as three or four a day for weeks on end. They were looking for something special, but we did all our real work well after sundown.

We called ourselves "The Nighthawks," which became the official nickname of our 37th Tactical Fighter Wing. For years on end we were forced to live like vampire bats in a dark cave. We slept all day behind thick blackout curtains and began to stir only when the sun went down. The F-117A is a night attack plane using no radio, no radar, and no lights. The Skunk Works stripped the fighter of every electronic device that could be picked up by ground-to-air defenses. The engines were muffled to eliminate noise. We flew below thirty thousand feet to avoid contrails on moonlit nights. We carried no guns or air-to-air missiles because the airplane wasn't designed for high-performance maneuvering, but to slip inside hostile territory, drop its two bombs, and get the hell out of there. So nights were meant for stealth, and we spent five nights a week practicing bombing runs and air-to-air refueling above the remote test range. We started work two hours after sunset and finished two hours before sunrise. Whenever the airplane left the hangar, the hangar lights had to be turned off. No landing field lights were allowed.

Our families had no idea about where we took off to every Monday or where we returned from every Friday evening. Most of us were family men who lived on base housing at Nellis, just outside Las Vegas. Going home on weekends by charter flights into Nellis was rough because we led normal lives for two days with our wives and kids before reverting to night-stalking vampires again. Marriages were really put to a severe test. In cases of emergency, wives would call a special number at Nellis and ask us to call home.

I know it sounds corny, but our morale stayed high because our task was to keep twelve airplanes on standby alert to go to war on the instant command of the president of the United States. Only the president or his secretary of defense could unleash us. And the second reason our morale stayed high was the airplane itself. All of us who flew it got to fall in love. We all agreed that if we flew within the assigned mission and stayed within the flight envelope, the stealth fighter was a sweetheart. Absolutely superb. And we all became proficient using smart and precise laser-guided bombs. We carried a pair of two-thousand-pounders that would follow our laser guide beam right into the heart of a target as we lined it up on crosshairs on our cabin video screen. We could find Mrs. Smith's rooming house and take out the northeast corner guest room above the garage. That kind of precision was awesome to behold.

I was made colonel in early 1990 and by mid-summer became the wing commander, just in time for our early August deployment to Saudi Arabia. And a few months after that we struck the first blow in Operation Desert Storm.

A year after the stealth fighter became operational, two computer wizards who worked in our threat analysis section came to me with a fascinating proposition: "Ben, why don't we make the stealth fighter automated from takeoff to attack and return? We can plan the entire mission on computers, transfer it onto a cassette that the pilot loads into his onboard computers, that will route him to the target and back and leave all the driving to us."

To my amazement they actually developed this automated program in only 120 days and at a cost of only $2.5 million. It was so advanced over any other program that the Air Force bought it for use in all their attack airplanes.

At the heart of the system were two powerful computers that detailed every aspect of a mission, upgraded with the latest satellite-acquired intelligence so that the plan routes a pilot around the most dangerous enemy radar and missile locations. When the cassette was loaded into the airplane's system, it permitted "hands-off" flying through all turning points, altitude changes, and airspeed adjustments. Incredibly, the computer program actually turned the fighter at certain angles to maximize its stealthiness to the ground at dangerous moments during a mission, when it would be in range of enemy missiles, and got the pilot over his target after a thousand-mile trip with split-second precision. Once over the target, a pilot could override the computers, take control, and guide his two bombs to target by infrared video imagery. Otherwise, our autopiloted computer was programmed even to drop his bombs for him.

It took us about two years to really perfect this system, aided by the nightly training flights at Tonopah. The computerized auto-system was so effective that on a typical training flight pilots were targeting particular apartments in a Cleveland high-rise or a boathouse at the edge of some remote Wisconsin lake and scoring perfect simulated strikes.

The first chance to test the airplane and the system under

real combat conditions came in April 1986, when the squadron
received top secret orders directly from Caspar Weinberger,
Reagan's defense secretary, to be prepared for a Delta Force–
style nighttime strike against Libya's Muammar Kaddafi's
headquarters in Tripoli. The mission was code-named Operation El Dorado Canyon and would involve eight to twelve
F-117As. Preparations were immediately made to fly to the east
coast, overnight for crew rest, then take off the next day and
using air-to-air refueling fly straight to Libya and hit Kaddafi
around three in the morning. Senior officers at the Tactical Air
Command who had been briefed about the existence of the
stealth fighter and were monitoring the training program underway at Tonopah advised Weinberger that the F-117A was
the perfect weapons system for this covert surgical strike operation. "This was why the system was built," one four-star declared. But Cap hesitated, and within an hour of the squadron's
scheduled departure to the east coast he scrubbed us from the
mission. He simply did not want to reveal the existence of this
top secret revolutionary airplane to the Russians that soon. In
my view Weinberger booted it. The raid was carried out using
Navy fighter-bombers off carriers, and Kaddafi escaped with
his life because Libyan defenses picked up the attackers coming in time to sound the alarm and several bombs aimed directly at Kaddafi's quarters missed their target because the
attacking aircraft were forced to evade incoming missiles and
flak. The F-117A would have attacked with surprise and placed
that smart bomb right on the guy's pillow.

The Defense Department reluctantly revealed the stealth
fighter's existence in 1988. The time had come to expand training operations to include other Air Force units, and the Pentagon intelligence analysts concluded that the Soviets already
knew the airplane was in the inventory. Although the press
had speculated about the existence of a stealth fighter for
years, what it actually looked like — its crucial shape and
design — remained safely secret. The press even called it the

F-19, the wrong designation, and published speculative artist's renditions that caused our experts like Denys Overholser and Dick Cantrell to laugh in glee. Still, I knew several high-level intelligence officials who were miffed that the Air Force officially unveiled the airplane at all. "The Russians," they told me, "are worried and puzzled. They don't have a clue about how to counter the F-117A. We've got them burning lights and working weekends. Much better, though, if we kept it under wraps until we hit them with it."

Other Voices
General Larry D. Welch
(Air Force Chief of Staff 1986–1990)

As Air Force Chief of Staff, what I had in mind for the F-117A was a specific set of extremely high-value targets that would neutralize the enemy defenses for a full-scale attack. For example, we had pondered endlessly about how we could cope with the Soviet SA-5 and SA-10 ground-to-air missiles — that is, by what means could we take them out and allow our air armada to proceed safely to their targets inside the Soviet heartland? We finally determined that the stealth airplane was our ticket. If we had a squadron of these revolutionary airplanes that no one knew about, and if they could take out those damned SA-5s, that gave us a tremendous strategic advantage over the Russians. As it turned out, we had a rather limited and myopic vision of what the airplane was really capable of. The conflict with Iraq proved it was far more versatile in undertaking all sorts of attack missions than any of us had ever imagined. Before the F-117A flew on the first night of combat in Operation Desert Storm, we had been forced to ponder how many days and sorties it would require before we could grind down enemy air defenses to the point where we could conduct a full-scale air campaign. The combination of stealth with its high-precision munitions provided an almost total assurance that

we could destroy enemy defenses from day one and the air
campaign could be swift and almost devoid of any losses. In
the past, you would have been betting your hat, ass, and spats
on a lot of wishful thinking to conceive a battle plan that would
eliminate most of the highest-value enemy targets over the
most heavily defended city on earth on the opening night of the
war. But that's exactly what Ben Rich's airplane did on the first
night of Desert Storm. For me, the shining moment came on
live television when that F-117A placed a bomb right down the
airshaft of the Iraqi defense ministry with the whole world
watching. Think about the impact that hit had on the entire
Iraqi leadership. And I'm certain that in every defense ministry
around the world there was an instant recognition that something astonishing had taken place with implications for future
air warfare that were impossible to imagine.

Donald Rice
(Secretary of the Air Force from 1989 to 1992)

I was at the Pentagon the night that Operation Desert Storm
kicked off. H-hour was scheduled to begin precisely at three
a.m. Baghdad time with precision raids staged by the F-117As.
The timing had to be exact and we had planned this opening
raid for weeks, so we were all disconcerted to suddenly see
CNN going live to Baghdad at twenty minutes to three to report that the city was under attack. There were three CNN reporters in a hotel room — Bernard Shaw, John Holliman, and
Peter Arnett — delivering excited accounts of cruise missiles
streaking past and sounds of attack airplanes overhead. The
sky was alive with tracers. This went on for twenty minutes,
during which not a thing was actually happening in the skies
over Baghdad — absolutely nothing.

With the exception of the F-117s, which had been sent
ahead and were already past the Iraqi border on their way to
attack Baghdad, the remainder of the allied air armada was
purposely being held back and out of range of a string of three

early-warning radar stations inside Iraq, near the Saudi border. We sent in Apache attack helicopters to take them out, and that attack was launched at twenty-one minutes to three. Apparently someone radioed back to Baghdad, "We're being attacked!" In Baghdad, they reacted by immediately firing everything they owned into the night skies. Finally, at one minute past three, one of the three CNN reporters said, "Whoops, the phone in our room just went dead." A minute later, at two minutes after three, the lights in that hotel room went out. That told us that the real attack had actually begun. We had preplanned that, at two minutes after three, the first F-117s would take out the telephone center and central power station in downtown Baghdad. And that's how we learned back in Washington that the leading elements of the F-117 attack force had dropped their precision bombs exactly on time.

We learned that night, and for many nights after that, that stealth combined with precision weapons constituted a quantum advance in air warfare. Ever since World War II, when radar systems first came into play, air warfare planners thought that surprise attacks were rendered null and void and thought in terms of large armadas to overwhelm the enemy and get a few attack aircraft through to do damage. Now we again think in small numbers and in staging surprise, surgically precise raids. Looking ahead, I'd predict that by the first couple of decades of the next century every military aircraft flying would be stealth. I might be wrong about the date but not about the dominance of stealth.

Colonel Barry Horne

Bats. Bats were the first visual proof I had that stealth really worked. We had deployed thirty-seven F-117As to the King Khalid Air Base, in a remote corner of Saudi Arabia, out of the range of Saddam's Scuds, about 900 miles from downtown Baghdad. The Saudis provided us with a first-class fighter base with reinforced hangars, and at night the bats would come out and feed off insects. In the mornings we'd find bat corpses littered around our airplanes inside the open hangars. Bats used a form of sonar to "see" at night, and they were crashing blindly into our low-radar-cross-section tails.

After all those years of training, we certainly believed in the product, but it was nice having that kind of visual confirmation, nevertheless. On the night of D-day in Desert Storm, it fell on us to hit first. Most of us felt like firefighters about to test a flame-retardant shield by walking into a wall of fire. The so-called experts assured us that the suit worked, but we really wouldn't know for sure until we made that fateful walk. As we suited up to fly into combat for the first time, one of the other pilots whispered to me, "Well, I sure hope to God that stealth shit really works."

He spoke for us all.

H-hour for Desert Storm was three a.m. Baghdad time, January 17, 1991. I climbed into my airplane shortly after midnight. Frankly, I don't think you could have driven a needle up my sphincter using a sledgehammer. From all our briefings we knew that we would be running up against the greatest concentration of triple A and missile ground fire since the Vietnam War, or maybe even in history. Saddam Hussein had sixteen thousand missiles and three thousand antiaircraft emplacements in and around Baghdad, more than the Russians had protecting Moscow. The F-117A was the only coalition airplane that would be used to hit Baghdad in this war. We got the missions most hazardous to a pilot's health. Otherwise the plan was to hit Saddam's capital with Navy Tomahawk missiles, fired from ships at sea.

Each of us carried two hardened laser-guided two-thousand-pounders designed to penetrate deep into enemy bunkers before exploding. They were called GBU-27s, and only the F-117As carried them. The mission took us five hours with three air-to-air refuelings. We came at Baghdad in two waves. Ten F-117As in the first wave, to knock out key communications centers, and then the second wave of twelve airplanes an hour or so later. The skies over Baghdad looked like three dozen Fourth of July celebrations rolled into one. Only it was a curtain of steel that represented blind firing. They could detect us, but they couldn't track us. We were like mosquitoes buzzing around their ears and they furiously swatted at us blindly. They just hoped for a golden BB — a lucky blind shot that would hit home, and I couldn't see how they could possibly miss. The only analogy I could think of was being on a ramp above an exploding popcorn factory and not having one kernel hit you. The law of averages alone would have made that impossible — and so I prayed.

That first night we saw French-built F-1s and Soviet MiG-29s flying around on our sensor displays. But they gave no sign of ever seeing us.

There were five communications sectors in the country, so we didn't have to destroy all their missiles or airplanes, but knock out their brains and claw out their eyes. So we hit their missile and communications centers, their operational commands, and their air defense center. In only three bombing raids that lasted a total of about twenty minutes, combined with attacks from Tomahawk missiles, we absolutely knocked Iraq out of the war. From that first night, they were incapable of launching retaliatory air strikes or sustaining any real defenses against our airpower. All they had left were mobile Scud missiles — a primitive revenge weapon — and vulnerable ground troops who had to fight in the open without air cover or hope. To put it in domestic terms, if Baghdad had been Washington, that first night we knocked out their White House, their Capitol building, their Pentagon, their CIA and FBI, took out their telephone and telegraph facilities, damaged Andrews, Langley, and Bolling air bases, and punched big holes in all their key Potomac River bridges. And that was just the first night. We went back night after night over the next month.

We flew in twos, but you don't see your partner, so the guys that first night saw the skies over Baghdad and just figured we'd lost airplanes and pilots. Then, once safely back across the border, we joined up and saw that everyone was okay and we were amazed, overjoyed, and deeply moved. No one had suffered as much as a hit or even a near miss. That stealth shit had really worked.

Major Miles Pound

It took only about two or three missions before most of us didn't even bother to glance out at the flak bursting all around us. We took advantage of their blind firing. We'd delay an attack five minutes knowing they'd have to stop firing to cool down their guns — then we'd come in and hit them. We drew all the demanding high-precision bombing of the most heavily defended, highest-priority targets. The powers that be decided to use B-52s to pour down bombs on the big North Taji military industrial complex. But it was protected by surface-to-air missiles that could knock down our bombing armada. So we went in the night before and took out all fifteen missile sites using ten stealth airplanes. They never saw us coming. That mission won us a standing ovation from General Schwarzkopf and the other brass monitoring us back in the coalition war room.

Given our precision bombs we could locate the one communications node in a city block and take it out without inflicting collateral damage. We used to brag, "Just tell us whether you want to hit the men's room or the ladies' room and we'll oblige." Because of stealth we could arrive at target unseen and focus entirely on making a precision hit. Our GBU-27 laser-guided bomb could penetrate the most hardened bunker. We hit Saddam's Simarra chemical bunkers with these bombs. They were eight feet of reinforced concrete, and we used the first bomb to pierce through the dense construction, then a second bomb followed down the same drilled hole made by the first and exploded with tremendous impact. About halfway through the war we began running low on bomb supplies and reverted to using a lighter bomb. Those GBU-10s bounced right off the roof of some hardened hangars at a forward Iraqi airbase called H-2. Intelligence reported the Iraqis were gleeful, felt they had finally defeated us at something, so they crammed these hangars with as many of their remaining jet fighters as they could. We waited for a couple of days, then went in with our heavier GBU-27s and blew that damned air base off the map.

Three other missions I remember with a lot of relish: we did some high-precision bombing and took out a Republican Guard barracks at a prison camp housing Kuwaiti prisoners, allowing them to escape. On another night, we took out Peter Arnett of CNN! I was at the base watching him broadcast — that guy was not wildly popular with many of us because we felt Saddam was just using him for propaganda — and we knew that in exactly six seconds our guys were going to hit the telecommunications center in downtown Baghdad and knock Arnett off the air. So we began counting out: "Five . . . Four . . . Three . . . Two . . . One." The screen went blank. Right on cue, too. We cheered like nuts at a football game.

But the raid against Saddam's nuclear research facility, which also had capability for chemical and biological weapons production, probably proved stealth at its best. The Air Force went after that place in daylight, using an armada of seventy-two airplanes, including fourteen attack F-16s, and the rest escorts, jammers, and tankers needed to support such an operation with conventional aircraft. Those pilots saw more SAMs and triple As coming up at them than they cared ever to remember. The Iraqis covered the target with smoke generators so that our guys had no choice but to drop bombs into the smoke and scoot for their lives. They scored no hits.

We came in at three in the morning using only eight airplanes and needing only two tankers to get us there and back, and took out three of the four nuclear reactors and heavily damaged the fourth. Once that first bomb hit all hell broke loose. I dropped my bombs, but I couldn't get my bomb-bay door closed. That was as bad as it could get because a right angle is like a spotlight to ground radar and a bomb-bay door is a perfect right angle. And out of the corner of my eye I saw a missile firing up at me. I had one hand on the eject lever and the other trying to manually close that stalled bomb bay. As the missile closed on me, the door finally did, too, and I watched that missile curve harmlessly by me as it lost me in its homing. About an hour later I began breathing again.

The night of that first raid against Baghdad coincided with a farewell banquet Lockheed staged to mark my retirement as head of the Skunk Works. It was a very emotional and patriotic evening, interspersed with the latest bulletins and live coverage from CNN. Early the next morning my son Michael called me and read me a story from the *New York Times* reporting that the first F-117A to drop a bomb on Baghdad carried a small American flag in its cockpit that later would be presented to me. The story said that the pilots of the 37th Tactical Fighter Wing had dedicated that first air strike to me in honor of my retirement. Even more gratifying was that stealth lived up to all our expectations and claims. In spite of undertaking the most dangerous missions of that war, not one F-117A was hit by enemy fire. I know that Colonel Whitley had privately estimated losses of 5 to 10 percent in the first month of the air campaign. No one expected to escape without any losses at all. The stealth fighters composed only 2 percent of the total allied air assets in action and they flew 1,271 missions — only 1 percent of the total coalition air sorties — but accounted for 40 percent of all damaged targets attacked and compiled a 75 percent direct-hit rate. The direct-hit rate was almost as boggling as the no-casualty rate since laser-guided bombs are strictly line of sight, depending on good visibility, and the air war was conducted during some of the worst weather in the region in memory.

The airplane was used at first as a silver bullet against high-value targets. They dropped the first bombs and opened the door for everyone else by destroying the Iraqi communications network. Those attacks were shown to the American public on CNN, and the political impact was as great as the military. It showed we could go downtown at will and with the precision of threading the eye of a needle take out the enemy military command centers with terrific accuracy. Those bull's-eye shots kept the public's morale high and its backing secure. No shoot-downs; no prisoners; no hostages.

Gradually, stealth missions were broadened to include air bases and bridges. Bridges are the most difficult target to destroy unless hit in a precise spot with the right payload. To bring down some bridges in Vietnam, for example, took thousands of sorties. The F-117A knocked out thirty-nine of the forty-three bridges spanning the Tigris-Euphrates River — simply astounding.

Stealth opened a new frontier in air war, proving that night attacks were more effective and less dangerous than daylight raids, where aircraft can be seen by the eye as well as by electronics. But Operation Desert Storm also raised red-flag warnings about future air combat: one month seemed to be the logistical limit to air combat sorties. We didn't design our airplanes to fly five hours a day, every day, for a month or more. Pilot fatigue and a shortage of spare parts became a growing concern. We almost ran out of bombs, too. But the overriding fact of Desert Storm was that the only way the enemy knew the F-117A was in the sky above was when everything around him began blowing up.

5

HOW WE SKUNKS GOT OUR NAME

I FIRST SHOWED UP at Kelly Johnson's front door, in December 1954, as a twenty-nine-year-old thermodynamicist earning eighty-seven bucks a week. I had never before set foot inside the so-called Skunk Works, in Building 82, a barnlike airplane assembly facility next to the Burbank airport's main runway, where Kelly and his minions held forth in a warren of cramped offices, oblivious to the outside world. Everything about that operation was secret, even what building they were in. All I knew for sure was that Johnson had called over to the main plant, where I had been working for the past four years, and asked to borrow a thermodynamicist, preferably a smart one, to help him solve some unspecified problems. It was like a band leader calling over to the union hall to hire a xylophone player for a one-nighter.

My expertise was solving heat problems and designing inlet and exhaust ducts on airplane engines. In those years, Lockheed was booming, cranking out a new airplane every two years. I felt I was in on the ground floor of a golden age in aviation — the era of the jet airplane — and couldn't believe my good luck. As young and green as I was, I had already earned my very own patent for designing a Nichrome wire to wrap around and electrically heat the urine-elimination tube used on Navy patrol planes. Crewmen complained that on freezing winter days their penises were sticking painfully to the metal funnel. My design solved their problem and I'm sure

made me their unknown hero. Both my design and patent
were classified "Secret."

My input was far less dramatic working on America's first
supersonic jet fighter, the F-104 Starfighter, nicknamed by the
press "the missile with the man in it" in tribute to its blazing
Mach 2 speed. I helped design the inlet ducts on that, as well
as on our first military jet transport, the C-130, and on the
F-90. The latter was a stainless-steel jet fighter, capable of
pulling twelve-g loads during incredible dives and turns, but
was woefully underpowered since the engine originally designed for it was canceled by the Air Force for budgetary reasons. So the F-90 wound up serving the country by being
shipped to Ground Zero at the Nevada atomic test site at a
mock military base specially constructed to determine how
various structures and military equipment would withstand
an A-bomb explosion. The short answer was everything was
either vaporized or blown to pieces except for the F-90. Its
windshield was vaporized, its paint sand-blasted, but otherwise our steel airplane survived in one piece. That sucker was
built.

The projects at Lockheed were all big-ticket items, and
workrooms as big as convention halls were crammed with
endless rows of white-shirted draftsmen, working elbow to elbow, at drafting tables. We engineers sat elbow to elbow, too,
but in smaller rooms and a slightly less regimented atmosphere. We were the analytical experts, the elite of the plant,
who decreed sizes and shapes and told the draftsmen what to
draw. All of us were well aware that we worked for Chief Engineer Clarence "Kelly" Johnson, the living legend who had designed Lockheed's Electra and Lockheed's Constellation, the
two most famous commercial airliners in the world. All of us
had seen him rushing around in his untucked shirt, a
paunchy, middle-aged guy with a comical duck's waddle,
slicked-down white hair, and a belligerent jaw. He had a thick,
round nose and reminded me a lot of W. C. Fields, but without

the humor. Definitely without that. Johnson was all business
and had the reputation of an ogre who ate young, tender engineers for between-meal snacks. We peons viewed him with the
knee-knocking dread and awe of the almighty best described
in the Old Testament. The guy would just as soon fire you as
have to chew on you for some goof-up. Right or not, that was
the lowdown on Kelly Johnson. One day, in my second year on
the job, I looked up from my desk and found myself staring
right into the face of the chief engineer. I turned pale, then
crimson. Kelly was holding a drawing of an inlet I had designed. He was neither angry nor unkind while handing it
back to me. "It will be way too draggy, Rich, the way you designed this. It's about twenty percent too big. Refigure it."
Then he was gone. I spent the rest of the day refiguring and
discovered that the inlet was eighteen percent too big. Kelly
had figured it out in his head — by intuition or maybe just
experience? Either way, I was damned impressed.

In those days Kelly wore his chief engineer's hat until
around two in the afternoon, then drove off to the Skunk
Works, which was about half a mile down the road tucked
away inside the Lockheed complex, and spent the final two
hours of his workday doing his secret design and development
work. There were always plenty of rumors about what Kelly
was up to — designing atomic-powered bombers or rocket-
driven supersonic fighters. Supposedly, he had a dozen engineers working for him, and we in the main plant pitied those
guys who were under that brutal thumb.

Still, the truth was I welcomed the chance to get out of the
main plant for a while. Lockheed was very regimented and
bureaucratic, and by my fourth year on the payroll I felt stymied and creatively frustrated. I had a wife and a new baby
son to support, and my father-in-law, who admired my moxie,
was pushing me to take over his bakery-delicatessen, which
earned the family a very comfortable living. I had actually
given notice to Lockheed, but at the last moment changed my

mind: I loved building airplanes a lot more than baking bagels or curing corned beef.

So I was eager to experience Kelly's Skunk Works, even if I was only on loan to him for a few weeks. It never occurred to me that I had any chance at all to stay there permanently. I was well trained in my engineering specialty and actually had taught thermodynamics at UCLA before joining Lockheed. I was also a naturalized American citizen and intensely patriotic, and welcomed the chance to work on secret projects designed to defeat the Russians. I had plenty of self-assurance and figured that as long as I did a good job, Kelly Johnson would behave himself.

As fearsome as Kelly was supposed to be, I knew he would be a pushover compared to my own stern father, Isidore Rich, a British citizen who had been, until the outbreak of World War II, the superintendent of a hardwood lumber mill in Manila, the Philippines, where I was born and raised. The Riches were among the first Jewish families to settle in Manila, and after one of my paternal grandfather's business trips to Egypt, he brought back a snapshot of the beautiful young daughter of one of his Jewish customers to show to my bachelor father. My father was enchanted, and a flirtatious correspondence bloomed into a full-fledged romance; marriage followed a few years later. My mother, Annie, was a French citizen, born and raised in Alexandria, a brilliant linguist who spoke thirteen languages fluently, a free spirit who pampered me, as her second youngest among four sons and one daughter. Mother was the opposite of our authoritarian father, who ruled over us like a Biblical prophet and used a strap to enforce his commandments. We lived in a big house with lots of servants on my father's very modest, middle-class salary, enjoying a way of life that was patently colonial and exploitive of the locals, but wonderfully secure and languid as the tropical air itself. My parents dressed formally to dine at their club and play bridge. We raised twenty-three police dogs in our huge backyard that resembled a tropical rain forest. And in later years, I amused Kelly with the story of how I built my first airplane at the tender age of fourteen. An older cousin bought a Piper Cub from a local flying club and to his dismay discovered that it came in a dozen crates and that he had to assemble it himself. My brothers and I built it for him in our big backyard, and after weeks of hard labor, we discovered that the finished product was too big to fit through our front gate. We had to take off the wings — but still no go. Then the tail, and finally the landing gear. In the end, we recrated the damned thing and my cousin got his money back. A few years later, that same cousin barely survived the infamous Bataan Death March.

By then, my family and I were safe in Los Angeles, having fled the island only a few months before the Japanese attack on Pearl Harbor. As tough as it was starting life anew, none of us were complaining: my father's sister, who had weighed a hefty one hundred and fifty pounds before the war, emerged from a Japanese prison camp as a gaunt, eighty-pound skeleton.

During the war years, my father and I worked in a Los Angeles machine shop to keep the family going. I was able to start college only after the war ended and I was already twenty-one. So I gave up my dream of becoming a doctor, like my father's brother, who was a world authority on tropical medicine, and decided to become an engineer instead. I graduated from Berkeley in mechanical engineering in 1949, in the top twenty in a class of three thousand, and decided to go on for my master's at UCLA, specializing in both aerothermodynamics and dating sorority girls. By then I had met my future wife, a beautiful young fashion model named Faye Mayer, who had an incomprehensible weakness for skinny engineers who smoked pipes. We got married just in time for me to start job hunting in the middle of the painful postwar recession and discover that a UCLA hotshot with a master's degree was just another candidate for unemployment. But one of my professors tipped me off to an engineer job opening at Lockheed's Burbank plant. I was hired and worked under Bernie Messinger, Lockheed's heat transfer specialist, when Kelly had phoned Bernie asking him to borrow a competent thermodynamicist for an undisclosed Skunk Works project. Bernie tapped me.

Since its inception back in 1943, when the first German jet fighters appeared in the air war over Europe, the Skunk Works had been entirely Kelly's domain. The War Department had turned to Lockheed's then thirty-three-year-old chief engineer to build a jet fighter prototype because he had designed and built the twin-engine P-38 Lightning, the most maneuverable propeller-driven fighter of the war. Kelly was given only 180 days to build that jet prototype, designed to fly at 600 mph, at least 200 mph faster than the P-38, at the very edge of the speed of sound. Kelly set to work by borrowing twenty-three of the best available design engineers and about thirty shop mechanics at the main plant. They operated under strict wartime secrecy, so that when he discovered that all available floor space in the Lockheed complex was taken for round-the-clock fighter and bomber production, that suited Johnson just fine. He rented a big circus tent and set up shop next to a noxious plastics factory, whose stench kept the curious at bay.

Around the time Kelly's crew raised their circus tent, cartoonist Al Capp introduced Injun Joe and his backwoods still into his "L'il Abner" comic strip. Ol' Joe tossed worn shoes and dead skunk into his smoldering vat to make "kickapoo joy juice." Capp named the outdoor still "the skonk works." The connection was apparent to those inside Kelly's circus tent forced to suffer the plastic factory's stink. One day, one of the engineers showed up for work wearing a civil defense gas mask as a gag, and a designer named Irv Culver picked up a ringing phone and announced, "Skonk Works." Kelly overheard him and chewed out Irv for ridicule: "Culver, you're

fired," Kelly roared. "Get your ass out of my tent." Kelly fired
guys all the time without meaning it. Irv Culver showed up for
work the next day and Kelly never said a word.

Behind his back, all of Kelly's workers began referring to
the operation as "the skonk works," and soon everyone at the
main plant was calling it that too. When the wind was right,
they could smell that "skonk."*

And who knows — maybe it was that smell that spurred
Kelly's guys to build Lulu Belle, their nickname for the cigar-
shaped prototype of the P-80 Shooting Star, in only 143
days — 37 days ahead of schedule. The war ended in Europe
before the P-80 could prove itself there. But Lockheed built
nearly nine thousand over the next five years, and during the
Korean War the P-80 won the first all-jet dogfight, shooting
down a Soviet MiG-15 in the skies above North Korea.

That primitive Skunk Works operation set the standards for
what followed. The project was highly secret, very high priority, and time was of the essence. The Air Corps had cooperated
to meet all of Kelly's needs and then got out of his way. Only
two officers were authorized to peek inside Kelly's tent flaps.
Lockheed's management agreed that Kelly could keep his tiny
research and development operation running — the first in
the aviation industry — as long as it was kept on a shoestring
budget and didn't distract the chief engineer from his principal duties. So Kelly and a handful of bright young designers
he selected took over some empty space in Building 82; Kelly
dropped by for an hour or two every day before going home.
Those guys brainstormed what-if questions about the future
needs of commercial and military aircraft, and if one of their
ideas resulted in a contract to build an experimental proto-

* In 1960, Capp's publisher objected to our use of Skonk Works, so we
changed it to Skunk Works and registered the name and logo as trademarks.
It is listed in the *Random House Dictionary:* "Often a secret experimental
division, laboratory or project for producing innovative design or products
in the computer or aerospace field."

type, Kelly would borrow the best people he could find in the main plant to get the job done. That way the overhead was kept low and the financial risks to the company stayed small.

Fortunately for Kelly, the risks stayed small because his first two development projects following the P-80 were absolute clunkers. He designed and built a prototype for a small, low-cost-per-mile transport airplane called the Saturn that was really a sixth toe on commercial aviation's foot because the airlines were buying the cheap war-surplus C-47 cargo plane to haul their customers and were calling it the DC-3. Then Kelly and his little band of brainstormers designed the damnedest airplane ever seen — the XFV-1, a vertical riser to test the feasibility of vertical takeoff and landing from the deck of a ship. The big trouble, impossible to overcome, was that the pilot was forced to look straight up at the sky at the crucial moment when his airplane was landing on deck. Even Kelly had to concede the unsolvability of that one.

But the open secret in our company was that the chief engineer walked on water in the adoring eyes of CEO Robert Gross. Back in 1932, Gross had purchased Lockheed out of bankruptcy for forty grand and staked the company's survival on the development of a twin-engine commercial transport. Models of the design were sent to the wind tunnel labs at the University of Michigan, where a young engineering student named Clarence Johnson contradicted the positive findings of his faculty advisers, who praised the design to Lockheed's engineering team. Johnson, all of twenty-three, warned Lockheed's chief engineer at the time that the design was inherently directionally unstable, especially with one engine out.

Lockheed was sufficiently impressed to hire the presumptuous young engineer, and learned quickly why this son of Swedish immigrants was nicknamed "Kelly" by his school chums years earlier. He might be stubborn as a Swede, but his temper was definitely Old Sod.

Kelly solved the Electra instability problem with an unconventional twin-tail arrangement that soon became his and Lockheed's trademark. The Electra revolutionized commercial aviation in the 1930s. Meanwhile Kelly was the shining light in the company's six-man aviation department — the expert aerodynamicist, stress analyst, weight expert, wind tunnel and flight test engineer — and he did some test flying himself. He once said that unless he had the hell scared out of him once a year in a cockpit he wouldn't have the proper perspective to design airplanes. Once that guy made up his mind to do something he was as relentless as a bowling ball heading toward a ten-pin strike. With his chili-pepper temperament, he was poison to any bureaucrat, a disaster to ass-coverers, excuse-makers, or fault-finders. Hall Hibbard, who was Kelly's first boss at Lockheed, watched Kelly work for three days during the war to transform Lockheed's Electra into a bomber for the British called the Hudson. The transformation was so successful that the RAF ordered three thousand airplanes, and Hibbard was so awestruck by Johnson's design skills, he claimed "that damned Swede can actually *see* air." Kelly later told me that Hibbard's remark was the greatest compliment he had ever received.

The Skunk Works was always strictly off-limits to any outsider. I had no idea who even worked there when I reported in that first day, just before Christmas, 1954, to Building 82, which was an old bomber production hangar left over from World War II days. The office space allocated to Kelly's Skunk Works operation was a narrow hallway off the main production floor, crowded with drilling machines and presses, small parts assemblies, and the large assembly area which served as the production line. There were two floors of surprisingly primitive and overcrowded offices where about fifty designers and engineers were jammed together behind as many desks as a moderate-size room could unreasonably hold. Space was at a premium, so much so that Kelly's ten-person procurement department operated from a small balcony looking down on the production floor. The place was airless and gloomy and had the look of a temporary campaign headquarters where all the chairs and desks were rented and disappeared the day after the vote. But there was no sense of imminent eviction apparent inside Kelly's Skunk Works. His small group were all young and high-spirited, who thought nothing of working out of a phone booth, if necessary, as long as they were designing and building airplanes. Added to the eccentric flavor of the place was the fact that when the hangar doors were opened, birds would fly up the stairwell and swoop around drawing boards and dive-bomb our heads, after knocking themselves silly against the permanently sealed and blacked-out windows, which Kelly insisted upon for security. Our little feathered friends were a real nuisance, but Kelly couldn't care less. All that mattered to him was our proximity to the production floor. A stone's throw was too far away; he wanted us only steps away from the shop workers, to make quick structural or parts changes or answer any of their questions. All the workers had been personally recruited by Kelly from the main plant and were veterans who had worked with him before on other projects.

The engineers dressed very informally — no suits or ties — because being stashed away, no one in authority except Kelly ever saw them anyway. "We don't dress up for each other," Kelly's assistant, Dick Boehme, told me with a laugh. I asked Dick how long I could expect to stay. He shrugged. "I don't know exactly what Kelly has in mind for you to do, but I'd guess anywhere from six weeks to six months."

He was slightly off: I stayed for thirty-six years.

Twenty designers were stashed away in choking work rooms up on the second floor. The windows were sealed shut, and in those days nearly everyone smoked. To my delight, I was sharing an office with only six other engineers composing the analytical section, most of whom I discovered I already knew from my previous work on the F-104 Starfighter. Without exception, these were all colleagues whom I had particularly admired at the time, so I gave Kelly a quick "A" for sharp recruiting, myself excepted of course. We were only two doors removed from the boss's big corner office.

Before I really got to work, Boehme handed me a piece of paper on which was mimeoed Kelly's "riot act" — ten basic rules we worked by. A few of them: "There shall be only one object: to get a good airplane built on time." "Engineers shall always work within a stone's throw of the airplane being built." "Any cause for delay shall be immediately reported to C. L. Johnson in writing by the person anticipating the delay." "Special parts or materials shall be avoided whenever possible. Parts from stock shall be used even at the expense of added weight. Otherwise the chances of delay are too great." "Everything possible will be done to save time."

"For as long as you work here, this is your gospel," Dick said. Then he told me we were working with the folks at Pratt & Whitney to modify a regular jet engine to fly higher by at least fifteen thousand feet than any airplane had ever before flown. There were some inlet problems that I would be addressing. I knew the Russians were mediocre engine builders, at least a generation behind us. I figured we were building a radically new high-flying long-range bomber. But then I was shown a drawing of the airplane and I let out a whistle of surprise. The wings were more than eighty feet long. It looked like a glider.

"What is that thing?" I exclaimed.

"The U-2," Boehme whispered and put a finger to his lips. "You've just had a look at the most secret project in the free world."

6

PICTURE POSTCARDS FOR IKE

THE FULL WEIGHT of government secrecy fell on me like a sack of cement that first day inside Kelly Johnson's guarded domain. Learning an absolutely momentous national security secret just took my breath away, and I left work bursting with both pride and energy to be on the inside of a project so special and closely held, but also nervous about the burdens it would impose on my life.

I hadn't been inside the Skunk Works two minutes before realizing that everything that happened there revolved around one man — Clarence L. "Kelly" Johnson. Kelly's assistant, Dick Boehme, wasn't about to brief me or tell me my duties. That was up to the boss himself, and Dick dutifully escorted me down the hall to Kelly's corner office and stood by while Kelly, in shirtsleeves behind a behemoth-size mahogany desk, half hidden by an impressive stack of blueprints, welcomed me with neither a smile nor a handshake, but got to the point immediately: "Rich, this project is so secret that you may have a six-month to one-year hole in your résumé that can never be filled in. Whatever you learn, see, and hear for as long as you work inside this building stays forever inside this building. Is that clear? You'll tell no one about what we do or what you do — not your wife, your mother, your brother, your girlfriend, your priest, or your CPA. You got that straight?"

"Yes, sure," I replied.

"Okay, first read over this briefing disclosure form which says what I've just said, only in governmentese. Just remember, having a big mouth will cost you twenty years in Leavenworth, minimum. Sign it and then we'll talk."

He then continued, "I'm going to tell you what you need to know so that you can do your job. Nothing more, nothing less. We are building a very special airplane that will fly at least fifteen thousand feet higher than any Russian fighter or missile, so it will be able to fly across all of Russia, hopefully undetected, and send back beautiful picture postcards to Ike."

I gulped.

"That's its mission. Edwin Land, who designed the Polaroid camera, is also designing our cameras, the highest-resolution camera in the world. He's got Jim Baker, the Harvard astronomer, doing a thirty-six-inch folded optic lens for us. We'll be able to read license plates. And we've got Eastman Kodak developing a special thin film that comes in thirty-six-hundred-foot rolls, so we won't run out."

He handed me a large folder crammed with papers. "I've got a guy working on the engine inlets and exit designs. Here's his work so far. I want you to review it carefully because I don't think he's up to speed. I also want you to take over all the calculations on what we'll need for cabin heating and cooling, hydraulics, and fuel control. I don't know how long I'll need you here: maybe six weeks. Maybe six months. I've promised to have this prototype flying in six months. That will mean working six-day weeks. At least."

He dismissed me with the back of his hand, and a few minutes later I was squeezed into an empty desk in a room jampacked with thirty-five growling, snorting designers and engineers, many of whom I had worked with on the F-104 Starfighter. Dick Fuller, an aerodynamicist who had come over from the main plant only the day before, was seated on one side of me, and a stability and control specialist named

Don Nelson was on the other side. Our desks touched. We could put our arms around each other without even stretching. In Kelly's tight little island, there was a yawning chasm between secrecy and privacy.

That first night I got home two hours later than usual, and Faye was not exactly delighted to see me. She had bathed our two-year-old son, Michael, put him to bed, and had eaten alone. Up until now I had always been able to share my day with her, and she enjoyed hearing about office gossip and some of the airplanes I worked on, even though I spared her the eye-glazing technical details of my work. I was one engineer who knew how damned boring other engineers could be when we talked shop at parties.

"Well, how did it go?" Faye asked.

I sighed. "From now on I'll be home closer to midnight than dinnertime and I have to work Saturdays. It's so secret we don't even have secretaries or janitors."

"Oh, my God," Faye exclaimed, "don't tell me you're involved with The Bomb!"

Five years later, when Francis Gary Powers was shot down over the Soviet Union, and the U-2 spy plane was revealed in headlines around the world, I was finally able to tell my wife that I helped build that airplane. "I figured as much," she insisted.

I erroneously assumed that we were building this U-2 for the U.S. Air Force. This assumption was based on the fact that the spy plane in question had wings and flew, and therefore would be in the province of blue-suiters. But Ed Baldwin set me straight. "Baldy" was Kelly's structural designer, as crusty as a pumpernickel, who would remain unmellowed more than twenty years later while working for me on the stealth fighter. Over lunch, I remarked at the absence of a single Air Force project officer on hand to monitor progress or kibitz as we built their airplane.

"Friend, this project is Central Intelligence Agency all the way," Baldwin remarked. "Everything about it is under the spook's direction."

"You mean the CIA will have its own air force?"

"You said it," Baldy grinned. "The rumor is that Kelly will give them all Lockheed test pilots to fly this thing. We're also going to furnish all their mechanics and ground crew and build them a training base somewhere out in the boonies."

As it turned out, Baldy's rumors were two-thirds accurate. The agency hired its own pilots from the ranks of the Air Force, but we put them on the books as Lockheed employees so that their payment came out of a special Lockheed account of laundered CIA money rather than straight government checks. The subterfuge was that the pilots were Lockheed employees involved in a government-contracted high-altitude weather and performance study.

Everything about this project was dark alley, cloak and dagger. Even the way they financed the operation was highly unconventional: using secret contingency funds, they backdoored payment to Lockheed by writing personal checks to Kelly for more than a million bucks as start-up costs. The checks arrived by regular mail at his Encino home, which had to be the wildest government payout in history. Johnson could have absconded with the dough and taken off on a one-way ticket to Tahiti. He banked the funds through a phony company called "C & J Engineering," the "C & J" standing for Clarence Johnson. Even our drawings bore the logo "C & J" — the word "Lockheed" never appeared. We used a mail drop out at Sunland, a remote locale in the San Fernando Valley, for suppliers to send us parts. The local postmaster got curious about all the crates and boxes piling up in his bins and looked up "C & J" in the phone book and, of course, found nothing. So he decided to have one of his inspectors follow our unmarked van as it traveled back to Burbank. Our security people nabbed him just outside the plant and had him signing national security secrecy forms until he pleaded writer's cramp.

Clearly, building this airplane was deadly serious business. Inside the Skunk Works our irreverent group privately scoffed at the "secret agent" mentality of the agency security guys who made us take aliases if we had to travel on business in connection with the project. I chose the name "Ben Dover," as in "bend over," the name of a British music hall variety star of my father's misspent youth. Still, all of us involved in building this particular airplane felt the weightiness of our mission. Kelly was regularly briefed at the agency on the real state of the world, which, he assured us, was 70 percent worse than anything we read in our morning papers. He didn't hide from us his view that the success of U-2 operations might make the difference between our country's survival or not.

The Russians were crashing development of an intercontinental ballistic missile with powerful liquid-fuel engines, and East-West tensions were strained to the breaking point. Both the United States and Soviet Union had already successfully tested H-bombs within the past year and seemed poised to use them. John Foster Dulles, Eisenhower's secretary of state, warned that we would go to the brink of war to combat Communist expansion, coining the term "brinkmanship" for his eye-to-eye confrontation technique. He acknowledged the Russians' incredible conventional might: they out-divisioned us by a factor of ten, out-tanked us by a factor of eight, out-airplaned us by a factor of four. But Dulles drew lines in the sand around the world and served notice that if the Communists crossed any one of them it would mean instant nuclear retaliation on a massive scale. "Going to the verge of war without actually getting into war is the necessary art," Dulles claimed. But that Russian bear seemed fifteen feet tall.

Around the time we started working three shifts getting the U-2 built in the summer of 1955, a national poll of adult Americans indicated that more than half the population thought it more likely that they would die in a thermonuclear war than of old-age diseases. Around the country some anxious people began digging fallout shelters in their backyards and stocking them with Geiger counters and oxygen tanks. They were motivated to keep on digging by front-page diagrams in their papers showing how an H-bomb exploded over Manhattan would trigger a four-mile fireball, vaporizing all in its path from Central Park to Washington Square, and creating more than a million casualties in less than two minutes.

Like millions of other couples raising a family, my wife and I were forced to ponder the unthinkable: what would we do if L.A. was nuked? Assuming we survived the blast, where would we go? How would we protect ourselves from radiation and fallout? To even raise such questions was heartbreaking because there were no answers apparent at all.

So I had no trouble motivating myself to work hard, long hours to build the U-2. This was the airplane our government was impatiently waiting for to breach the Iron Curtain and finally discover the scope and dimensions of the Soviet threat. There was no way to hide from our cameras. And there was no hostile action the Russians could take that could stop us from our flights. We would be flying beyond reach of their defenses.

Eisenhower was being regularly briefed on our progress and sent word to Kelly via John Foster's brother, Allen Dulles, who ran the CIA, to crack the whip and get that U-2 launched. The president was receiving persistent warnings from the Joint Chiefs and the CIA that the Russians might be preparing to launch a preemptive first-strike nuclear attack against the United States. The evidence was fragmentary but unsettling. Khrushchev had bragged, "We will bury you," and at the 1954 May Day parade he gave our military attachés a chilling glimpse at what seemed to be his latest grave-makers: a half dozen new long-range missiles on huge portable launchers being trucked through Red Square, while overhead wave after wave of a new heavy bomber rattled the Moscow rooftops. Our military observers from the embassy counted one hundred bombers, nicknamed the Bison, that were capable of reaching New York with a nuclear payload. Only after the first U-2 flights was this estimate reassessed and our observers realized they had probably been duped: the Russians appeared to have flown the same twenty or so bombers over the Kremlin in a big circle. But the missiles seemed to confirm spy reports that the Soviets were working on a huge 240,000-pound-thrust rocket engine. Why this crash program in long-range weapons?

We were countering the Bisons with a new bomber of our own, the long-range B-52, but the trouble was we didn't have enough reliable information on precise locations of Russian bases and key industries to devise strategic targeting plans. We possessed only the most rudimentary idea of where vital Soviet bases and industrial centers were located or how well they were defended, or the kinds of terrain that a bombing mission would encounter going in and coming out. A massive amount of photomapping and technical intelligence was needed to provide the Strategic Air Command with an up-to-date comprehensive targeting plan.

Without an airplane like the U-2 that could overfly at heights above harm, the blue-suiters were driven to use aggressive, dangerous tactics. Had the American public known about the ongoing "secret air war" between the two superpowers they would have been even more in despair than many already were about the state of the world. There had been dozens of American attempts during the early 1950s to gather important Russian radar and electronic communications frequencies by flying provocatively up against the Soviet coastline and occasionally overflying their territory by as much as two hundred miles. Several of these unarmed reconnaissance aircraft were shot down either by Soviet jets or ground fire.

Most of the crews, totaling more than a hundred servicemen, simply disappeared off the scope and were presumed to have been sent to Siberia and/or killed.

Eisenhower finally ordered fighter escorts for these reconnaissance missions, resulting in several fierce dogfights with Soviet MiGs over the Sea of Japan. Ike was normally very cautious, but he was so intent on gaining information on Soviet missile development that he approved a joint CIA–British air force operation in the summer of 1955, in which a stripped-down Canberra bomber flew at fifty-five thousand feet, well above the range of Soviet fighters, and photographed the secret missile test facility called Kapustin Yar, east of Volgograd. The Canberra was hit more than a dozen times by ground fire and barely made it back to base. The crew reported that the Soviets seemed to have been alerted to the mission, and years later the CIA concluded that the operation had indeed been compromised by the notorious Kim Philby, a high-level official in British intelligence, who was a mole for the KGB.

In a final act of desperation before our spy plane could be launched, the blue-suiters began sending up spy balloons over Russia loaded with electronic gathering devices. They were announced as a weather systems survey, but the Soviets weren't fooled and immediately fired off angry protests to Washington. They also shot down some of the balloons, while the majority floated off into limbo. Only about thirty made it back to our side, and we actually learned a lot of useful information about Russian weather, especially wind patterns and barometric pressures.

This was pathetic, primitive stuff compared to the promise of our U-2. Dr. Edwin Land, who had pushed the idea of a high-flying spy plane in his role as a special technical consultant to the White House, had promised President Eisenhower a tremendous intelligence bonanza: "A single mission in clear weather can photograph in revealing detail a strip of Russia two hundred miles wide and twenty-five hundred miles long and produce four thousand sharp pictures," he wrote in his proposal. Land predicted the U-2 would obtain a detailed photographic record of Soviet railroads, power grids, industrial facilities, nuclear plants, shipyards, air bases, missile test sites, and any other target of strategic value. "If we are successful, it can be the greatest intelligence coup in history," Land assured the president.

We had stretched the design of this airplane to the limit to achieve unprecedented range and altitude. It could fly nine hours, travel four thousand miles, and reach heights above seventy thousand feet. The wings extended eighty feet, providing unusual lift capacity, like a giant condor gliding on the thermals, except, of course, that the U-2 did no gliding and flew high above the jet stream. Our long wings stored 1,350 gallons of fuel in four separate tanks.

Each pound adding to the airplane's overall weight cost us one foot of altitude, so while building the U-2 we were ruthless weight-watchers. Seventy thousand feet was our operational goal. Intelligence experts believed (erroneously as it turned out) that that altitude put our pilots beyond the range limits of Soviet defensive radar. That height was, however, beyond reach of their fighters and missiles.

We designed and built that airplane for lightness. The wings, for example, weighed only four pounds per square foot, one-third the weight of conventional jet aircraft wings. For taxiing and takeoffs, jettisonable twin-wheeled "pogos" were fitted beneath the enormous fuel-loaded wings and kept them from sagging onto the runway while taking off. The pogos dropped away as the U-2 became airborne.

The fuselage was fifty feet long, built of wafer-thin aluminum. One day on the assembly floor, I saw a worker accidentally bang his toolbox against the airplane and cause a four-inch dent! We looked at each other and shared the same unspoken thought: was this airplane too damned fragile to fly? It was a fear widely shared inside the Skunk Works that quickly transferred onto the flight line. Pilots were scared to death flying those big flapping wings into bad weather situations — afraid the wings would snap off. The U-2 had to be handled carefully, but proved to be a much tougher, more resilient bird than, frankly, I would ever have guessed. The landing gear was the lightest ever designed — weighing only two hundred pounds. It was a two-wheel bicycle configuration with a nose wheel and a second wheel in the belly of the airplane. Tandem wheels were used on gliders, but this was the first time ever for a powered airplane, which usually had tricycle landing gears. Ours would cause pilot trepidations about landing the U-2 that never quite evaporated, no matter how many landings a pilot successfully completed. Adding to the sense of the airplane's fragility was that the razor-thin tail would be attached to the fuselage by just three five-eighth-inch bolts.

The heart of the U-2 were hatches in the equipment, or Q, bay that would house two high-resolution cameras, one a special long-focal-length spotting camera able to resolve objects two to three feet across from a height of seventy thousand feet, and the other a tracking camera that would produce a continuous strip of film of the whole flight path. The two cameras weighed 750 pounds. Kelly and Dr. Land argued constantly about each other's needs to dominate the relatively small space inside those bays. Kelly needed room for batteries; Land needed all the room he could get for his bulky folding cameras. Kelly's temper flashed at Land: "Let me remind you, unless we can fly this thing, you've got nothing to take pictures of." In the end they compromised.

My principal work was on the engine's air intake, which had to be designed and constructed with absolute precision to maximize delivery of the thin-altitude air into the compressor face. Up where the U-2 aimed to cruise, just south of the Pearly Gates, the air was so thin that an oxygen molecule was about as precious as a raindrop on the Mojave desert. So the intakes had to be extremely efficient to suck in the maximum amount of oxygen-starved air for compression and burning. The real crunch was building a reliable engine for flying at the top of the stratosphere and finding special fuel that could operate effectively with so little oxygen. Pratt & Whitney built the highest-pressure-ratio engine available at that time, their J57 engine, which Kelly hoped could somehow be adopted for the U-2. He had met with Bill Gwinn, the head of Pratt & Whitney, at the company's main plant in Hartford, Connecticut.

"Bill," he said, "I need to fly at seventy thousand feet." Gwinn scratched his head. "We've never come close to that height, Kelly. I have no idea what's the fuel consumption and thrust needed to get up that high."

He put his best people to work on the problem. They were modifying most of the J57's innards — the alternator, oil cooler, hydraulic pump, and other key parts for extreme-altitude flying. The two-spool compressor and three-stage turbine were being hand-built. Even with these modifications, the engine would be able to produce only 7 percent of its take-off sea-level thrust at seventy thousand feet. The U-2 would be flying where outside temperatures would be minus 70 degrees F, causing standard military JP-4 kerosene fuel to freeze or boil off due to low atmospheric pressures. So Kelly turned to retired General Jimmy Doolittle, who was a key Eisenhower adviser on military and intelligence matters, as well as a board member of Shell Oil. Doolittle put the muscle on Shell to develop a special low-vapor kerosene for high altitudes. The fuel was designated LF-1A. The rumor about the LF abbreviation was that it stood for "lighter fluid." The stuff smelled like lighter fluid, but a match wouldn't light it. Actually, it was very similar in chemistry to a popular insecticide and bug spray of that era known as Flit. Once our airplane became operational, Shell diverted tens of thousands of gallons of Flit to make LF-1A in the summer of 1955, triggering a nationwide shortage of bug spray.

Kelly suffered stress headaches worrying about the engine

and fuel performance at such incredible altitudes. Several of our own engineers were dubious that a conventional jet engine could ever be made to function properly in a realm where experimental ramjets had flown for only minutes at blistering supersonic speeds. That kind of tremendous brute power was necessary to gulp down enormous quantities of oxygen-thin air. All of us worried about what would happen if the engine died above Russia, forcing the pilot to glide to lower altitudes to restart, placing him in range of Soviet missiles and fighters.

I had never before worked with so much intensity and camaraderie. Very quickly forty-five-hour workweeks would seem a luxury. We began logging sixty- to seventy-hour workweeks to meet the schedule. I had begun by reviewing the work of my predecessor, who I thought had done a competent job. But I quickly learned that Kelly had blind spots about certain people that could never be changed. For instance, I observed that he was particularly harsh in his dealings with a couple of engineers whom I considered to be extraordinarily good, and in my youthful naivete it never dawned on me that there might have been jealousy at play. Kelly was so hands-on that I quickly lost self-consciousness around him, although that was certainly not true of most others. I actually observed guys flushing and breaking out in a nervous sweat every time they had to deal with him — even several times a morning.

Very quickly I felt part of his team but far from being a key player. Some days he remembered my name and other times he clearly fudged it. But for whatever reason, I discovered that I really was not afraid of him. If I screwed up, I quickly admitted it and corrected my mistake. For example, one day I suggested something that would have added a hydraulic damper into the design and that meant decreasing altitude by increasing weight. I saw Kelly's face cloud over before I even finished speaking, and I immediately slapped my forehead and said, "Wait a minute. I'm a dumb shit. You're trying to take off pounds. . . . Back to the drawing board." The guys who tried to

finesse mistakes and hoped that Kelly would not notice usually wished they had never been born. Nothing got by the
boss. *Nothing*. And that was my sharpest impression of him,
one that never changed over the years: I had never known anyone so expert at every aspect of airplane design and building.
He was a great structures man, a great designer, a great aerodynamicist, a great weights man. He was so sharp and instinctive that he often took my breath away. I'd say to him, "Kelly,
the shock wave coming off this spike will hit the tail." He
would nod. "Yeah, the temperature there will be six hundred
degrees." I'd go back to my desk and spend two hours with a
calculator and come up with a figure of 614 degrees. Truly
amazing. Or, I'd remark, "Kelly, the structure load here will
be . . ." And he would interrupt and say, "About six point two
p.s.i." And I'd go back and do some complicated drudge work
and half an hour later reach a figure of 6.3.

Kelly just assumed that anyone he selected to work for him
would be more than merely competent. I assume he felt that
way about me. But during those feverish days of getting that
first U-2 prototype built, I was just another worker bee in his
swarming hive. And I actually learned to love our slumlike
working conditions. Everyone smoked in those days and the
smoke clouds resembled a thick London fog. Since no outsiders, including secretaries or janitors, were allowed near us,
we did our own sweeping up and took turns making our own
coffee. Working to giddiness, we acted like college sopho-
mores a shocking amount of the time. We hung "daring" pictures of Petty girls in scanty swimsuits which could be flipped
around to reveal on the opposite side a reproduction of water-
fowl. On rare occasions, when Kelly brought in visitors, someone would shout, "Present ducks," and we'd flip our three
framed pictures of full-breasted beauties. Once we had a contest to measure our asses with calipers. Leave it to me, I had
never won a contest before in my life and *I* won that one. I was
presented with a certificate proclaiming me "Broad Butt of the

Year." From then on, "Broad Butt" was my nickname. It was
still better than Dick Fuller's, though. Everyone called him
"Fulla Dick." And we were supposedly an elite group doing
momentous work.

The CIA was not at all in evidence unless I knew who I was
looking for. Every few weeks I would catch a glimpse of a tall,
patrician gentleman dressed improbably in tennis shoes,
freshly pressed gray trousers, and a garish big-checked sport
jacket that any racetrack bookmaker would have been proud
of. I once asked Dick Boehme who that guy was, and he replied with a stern "What guy? I don't see a soul." Kelly made
sure that few of us had any dealings with the visitor who appeared every few weeks or so. Many months went by before I
heard someone refer to him as "Mr. B." No one besides Kelly
knew his name. "Mr. B" was Richard Bissell, former Yale eco-
nomics professor and Allen Dulles's special assistant, put in
charge of running the CIA's spy plane project, who became the
unofficial godfather of the Skunk Works, the government official who really put us on the map. He became one of Kelly's
closest confidants and our most ardent champion. Ultimately,
he ran all the spy plane and satellite operations for the agency
until the last months of the Eisenhower administration in late
1959, when Allen Dulles put him in charge of organizing a
group of Cuban émigrées into a rag-tag battle brigade that
would attempt to invade the island at the Bay of Pigs. But in
the early days of the U-2, he was a mysterious figure to most of
us, part of a complicated working arrangement involving the
agency, Lockheed, and the Air Force that was unprecedented
in the annals of the military-industrial complex.

The operational plan for deploying our highly secret airplane was approved personally by President Eisenhower. Under this plan, the CIA was responsible for overseeing
production of the airplane and its cameras, for choosing the
bases and providing security, and for processing the film, no

mean feat since the special, tightly wound film developed by Eastman Kodak would stretch from Washington halfway to Baltimore on each mission. The Air Force would recruit the pilots, provide mission and weather planning, and run the daily base operations. Lockheed would design and build the airplane and provide ground crews for the bases and a cover for the pilots, who would carry Lockheed IDs and be officially logged on the company books as pilots for a government-contracted weather investigation program.

The reason why Kelly could move so quickly building the U-2 was that he could use the same tools from the prototype of the XF-104 fighter. The U-2, from nose to cockpit, was basically the front half of the F-104, but with an extended body from cockpit to tail. Using that tooling would save many months and a lot of money. Our goal was to put four birds in flight by the end of the first year. Each airplane would cost the American taxpayers $1 million, including all development costs, making it the greatest procurement bargain ever.

By April 1955, the first U-2 was being built under tight wraps inside the assembly area of Building 82, and Kelly sent for his chief engineering test pilot, Tony LeVier, who had flight-tested all of Johnson's airplanes since the days of the P-38. "Close the goddam door," he said to Tony. "Listen, you want to fly my new airplane?" Tony replied, "What is it?" Kelly shook his head. "I can't tell you — only if you say yes first. If not, get your ass out of here." Tony said yes. Kelly reached into his desk and unrolled a large blueprint drawing of the U-2. Tony began to laugh. "For chrissake, Kelly, first you have me flying your goddam F-104, which has the shortest wings ever built, and now you got me flying a big goddam sailplane with the longest wingspan I ever saw — like a goddam bridge."

Kelly rolled up the drawing. "Tony, this is top secret. What you just saw you must never ever mention to another living soul. Not your wife, your mother, nobody. You understand?

Now, listen. I want you to take the company Bonanza and find us a place out on the desert somewhere where we can test this thing in secret. And don't tell anyone what you're up to."

LeVier knew the vast sprawl of desert terrain shared by California and Nevada as well as any mule-packing Forty-Niner; as a test pilot he had mapped in his mind nearly every dry lake bed between Burbank and Las Vegas as a possible emergency landing strip. So he took off on his scouting expedition, after telling fellow pilots he was off to count whales for the Navy — a project Lockheed had actually done from time to time — and headed north toward Death Valley. Two days later, he found the perfect spot. "I gave it a ten plus," he told me years later. "Just dandy. A dry lake bed about three and a half miles around. I had some sixteen-pound cast-iron shotput balls with me and dropped one out to see if the surface was deep sand. Damned if it wasn't hard as a tabletop. I landed and took pictures." A few days later Tony flew Kelly and a tall civilian introduced to him only as "Mr. B." to the site to take a look. His wife had packed a picnic lunch, but a stiff wind began howling, blowing large stones across the surface of the dry lake. "This will do nicely," Mr. B. remarked. The area was not only remote but off-limits to all unauthorized air traffic because of its proximity to nuclear testing. As Kelly noted in his private log that day: "Flew out and located runway at south end of lake, then flew back (very illegally) over the atomic bomb sitting on its tower about nine hours before it was set to go off. Mr. Bissell pleased. He enjoyed my proposed name for the site as 'Paradise Ranch.'"

From mid-May to mid-July the pressure on the workers building the first U-2 grew in intensity to a point where three shifts were working eighty hours weekly. To put an airplane in the sky in only eight months was a tremendous achievement. On June 20, 1955, Kelly noted in his log: "A very busy time in that we have only 650 hours to airplane completion point. Having terrific struggle with the wing."

That long narrow wing was two-thirds as long as the length
of the fuselage and crucial to sustained high-altitude flight.
But wings that long created structural problems, including a
bending instability in flight known as aeroelastic divergence, a
fancy way of describing wings flapping like a seagull's and
possibly tearing off. We worked on the problem around the
clock. Kelly, meanwhile, was a blur of activity, juggling five or
six production problems simultaneously. As our airplane
neared completion, he was also sweating out the construction
of our remote facility. Fronting for the CIA under the phony
C & J Engineering logo, he hired a construction company to
put in wells, two hangars, an airstrip, and a mess hall in the
middle of a desert in blistering 130-degree summer heat. At
one point, the guy Kelly used as his contractor put out a subcontracting bid. One subcontractor warned him: "Look out
for this C & J outfit. We looked them up in Dun & Bradstreet,
and they don't even have a credit rating." This base was built
for only $800,000. "I'll bet this is one of the best deals the
government will ever get," Kelly remarked to several of us.
And he was right.

By early July both the airplane and the test site were near-
ing completion when Kelly suffered a nearly fatal car wreck,
after a driver ran a red light in Encino and clobbered him. He
was hospitalized with four broken ribs but hobbled back to
work in less than two weeks, just in time for what he referred
to in his log as "a terrific final drive to finish the airplane."

The first U-2 was completed on July 15, 1955. I remember
the sense of shock I experienced the first time I stood next to it
on the assembly floor. The airplane was so low slung that although I was slightly less than six feet tall, my own nose was
higher than the airplane's. Over the next few days, the airplane
was subjected to all kinds of flutter and vibration and control
tests, culminating in the most severe test of them all — Kelly's
personal final check and inspection. "I found thirty items to
improve," he told Dick Boehme with a grimace.

On July 23, the airplane was disassembled and loaded into special shipping containers. At four in the morning, the containers were loaded in a remote section of the Burbank airport onto a C-124 cargo plane and roared off before sunrise, headed for the desert base. Kelly followed in a C-47. We unloaded the bird on schedule into the semi-completed hangars and assembled it. We were ready to fly.

Other Voices
Tony LeVier

In early July, Kelly called me in and told me to get ready to go up to the "Ranch," as he called the base, and start flight tests. First time I flew there since the day I took Bissell and Kelly up, I almost fainted at the changes. Holy mackerel, they had put in a runway, had a control tower, two big hangars, a mess hall, a whole bunch of mobile homes. We had on hand only four engineers and twenty maintenance, supply, and administration people. Nowadays they'd probably use twenty people just to fuel an airplane.

The U-2 was very light, very fragile, very flimsy. Kelly wanted to know how I planned on landing it. I had never landed on glider wheels before — in tandem. Usually a pilot likes to make a landing approach nose high. But the landing problem was on Kelly's mind, causing him concern. I got advice from other pilots, who said not to land it on the nose wheel, otherwise I faced the danger of porpoising, which could lead to a structural breakup. But Kelly contradicted that advice. He said, "No. I want you to land it on the nose wheel. Otherwise, if you come in dragging your tail, nose high, I'm afraid you might stall out and lose the airplane totally."

Dry lake beds are very tricky to land on at times. Given desert lighting conditions, you can't always tell how low you are. So I had them lift the U-2 off the ground, so that the wheels were barely touching, just as if I was first touching down on a

landing, and the horizon out there was the horizon I would see
as I came in. I sat in the cockpit and I took black tape and
marked it on the cockpit glass even with the natural horizon. I
did that on both sides. The black tape markers would tell me
when I was lined up precisely with the horizon and that meant
my wheels were just touching the ground.

On August 2, 1955, I made my first taxi test in the airplane.
Towed it out on the lake three hundred feet. Kelly told me to
taxi and throttle up to fifty knots and then hit the brakes. I
pushed down on the pedals. God, they were sorry brakes. Kelly
got on the horn and said, "Okay, now take it up to sixty knots
and hit those brakes." I did as I was told. Then he said, "Now
take it up to seventy knots." So that's what I did, and I realized
we were suddenly in the goddam air. The lake bed was so
smooth I couldn't feel when the wheels were no longer touching. I almost crapped. Holy Christ, I jammed the goddam
power in. I got into stall buffet and had no idea where the goddam ground was. I just had to keep the goddam airplane under
control. I kept it straight and level and I hit the ground hard.
Wham! I heard thump, thump, thump. I blew both tires and
the damned brakes burst into flame right below the fuel lines.
The fire crew came roaring up with extinguishers followed by
Kelly in a jeep and boiling mad. "Goddam it, LeVier, what in
hell happened?" I said, "Kelly, the son of a bitch took off and I
didn't even know it." Who'd of guessed an airplane would take
off going only seventy knots? That's how light it was.

Our first real flight test took place late in the afternoon, a
few days later, on August 4. I took off around four in the afternoon, with big black thunderclouds building fast. I took her up
to eight thousand feet, with Kelly following behind me in a
T-33 piloted by my colleague Bob Mayte. I got on the horn:
"Kelly, it flies like a baby buggy." Rain was starting to splatter
the windshield, so we decided to cut the first flight short because of the weather. Kelly was getting edgy as I circled around
to make my approach for a landing. "Remember, I want you to

land it on the nose wheel." I said I would. I came down as gently as I knew how and just touched the nose wheel to the ground and the damned airplane began to porpoise. I immediately pulled up. "What's the matter?" Kelly radioed. The porpoising effect could break up that airplane — that was the matter. I told him I just touched the damned thing down and it began to porpoise on me. He told me, "Take it around and come in even lower than last time." I did that exactly and the damned thing started to porpoise again. I gunned it again. By now it's really starting to get black and the rain and wind are kicking up. Kelly is in full panic now. I can hear it in his voice. He's afraid the fragile airplane will come apart in the storm. He yells at me, "Bring it in on the belly." I say to him, "Kelly, I'm not gonna do that." I came around the third time and I held her nose high, just like I had wanted to, and put her down in a perfect two pointer, slick as a cat's ass. Bounced a little, but nice enough. The minute I was down, the sky opened up and it poured, flooding the lake bed under two inches of water. That night we had a big party and we all got smashed. "Tony, you did a great job today," Kelly said to me. Then he challenged me to an arm wrestle. The guy was strong as two oxen, but what the hell. He banged my arm down so hard he almost busted my wrist. I had it all bandaged up the next day. "What in hell happened to you?" he asked me. He was so soused he didn't even remember arm-wrestling me."

On that day British and West German intelligence finished tunneling into East Berlin to eavesdrop on Soviet and East German military headquarters. Allen Dulles visited the Oval Office and made his report personally to President Eisenhower: "I've come to tell you about two successes today — one very high and the other very low."

7

OVERFLYING RUSSIA

A MONTH after the first U-2 flight, the Skunk Works' test pilots were soaring 70,000 feet above the desert, breaking all existing altitude records in secret. After a few months, our pilots had logged 1,000 hours of flight time, had been to 74,500 feet, and had flown ten-hour 5,000-mile missions on one tank of gas.

Kelly was delighted by the airplane's performance even though our pilots experienced frequent engine stall-outs at these extreme altitudes, forcing Pratt & Whitney's engineers to log huge overtime adjusting their high-altitude engine to become more efficient.

With its enormous wingspan, designed to provide quick lift, the U-2 was able to glide for 250 miles from 70,000 feet, taking more than an hour to do so. Pilots couldn't restart their engine unless they descended to the more oxygen-rich altitude of 35,000 feet or lower. Meanwhile, that damned engine caused another big headache by spraying oil onto the cockpit windshield via the compressor that ran the cockpit air-conditioning. That was my domain. The airplane held sixty-four quarts of oil, and we often had to replace twenty lost quarts after a flight. Our pilots breathed potentially volatile pure oxygen inside their sealed helmets, while their windshield dripped potentially volatile hot oil. I tried all kinds of solutions, but in desperation I heeded a suggestion made by one of our veteran mechanics: "Why don't we just stuff Kotex around the oil filter and absorb the mess before it hits the windshield."

With *great* hesitation I approached the boss. Those steely eyes narrowed and he studied me hard. I saw my brief career at his side evaporate in one explosive bellow: "Rich, you're out of here!"

Kelly silently heard my sanitary napkin suggestion, then raised his eyebrows, shrugged, and said, "What the hell, give it a shot." I called the crew out at the facility and told them to stand by for a delivery of industrial-size cartons of sanitary napkins being airlifted their way immediately. And, by God, it worked!

But then a mysterious problem suddenly developed that held potentially disastrous consequences. The ground crews began reporting broken rubber seals inside engine valves and leaking pressure seals around the cockpit. The rubber had badly oxidized in only a few weeks, leaving all of us scratching our heads. We replaced the seals, but a few weeks later the seals leaked again. As it turned out, the answer to the mysterious malady was revealed one day on the front page of the *Los Angeles Times*, just beneath the fold. The article reported how European-made automobile tires were proving to be totally unacceptable for Los Angeles motorists. Because of our smog, the article reported, the rubber was badly oxidizing and causing "tire fatigue," leading to flats and rapid deterioration. The villain was ozone, a key component of our noxious smog. U.S. tire manufacturers, aware of the smog problem, added silicone to the rubber for their tires shipped to Southern California in order to avoid this oxidation problem. Reading that story, I almost jumped out of my chair. The U-2 was flying at the top of the troposphere, which was heavily laden with ozone. I mentioned the article to Dick Boehme, the U-2 program manager, who took it directly to Kelly. The fix was made quickly. All our seals were replaced with silicone and the problem vanished.

Despite the dreadful hours and the problems they caused in family life, the Skunk Works was for me far more splendid than a misery. Each day I found myself stretching on tiptoes to keep pace with my colleagues. Working with that crew was invigorating and fun. One of my favorites was our hydraulics guru, Dave Robertson, who in his spare time built toy square shells for a toy square cannon he invented, just to prove it could work. One Sunday I went over to his house and we lit the powder charge on the front lawn. Boom! The square projectile shot in a high arc across the street and blasted through the neighbor's upstairs window. "Wow," Dave grinned, "that little sucker really works!"

I turned to Dave for help and advice during that period of U-2 test flights in the summer of 1955, when our test pilots began reporting "duct rumble" at fifty thousand feet, describing the sensation as driving down a deeply rutted road on four uneven tires. In an airplane as fragile as the U-2, such severe shaking was a serious problem. The cause was flying at a slant so that more air was entering one of the twin air-intake ducts than the other. The problem landed directly in my lap since I had designed the intakes. Dave helped me design a splitter to enhance more even airflow and that helped to alleviate the problem, but not entirely. At fifty thousand feet, pilots were continuing to experience a roughness, although not to the point of watching their wing flaps so that they broke into a cold sweat. I told Kelly, "We've got it under control, but it won't go away. I have no idea why it happens only at fifty thousand feet." He didn't either. He just told our pilots: "Avoid flying at fifty thousand whenever possible. You should be up higher than that anyway." Pratt & Whitney finally solved the problem completely a year or so later by revising the fuel control for a better match of air and fuel into the engine.

But our test pilots had a lot more on their minds than rumbling ducts. Landing the U-2 on its two tandem wheels was neither easy nor routine. Our veteran test pilots warned Kelly that training CIA pilots to fly the U-2 and not getting one or more killed in the process was going to be a major challenge. Pilots were also apprehensive when hitting clear air turbulence and watching those long thin wings flapping like a bird's, worrying that the next big gust would snap them off entirely. And, by the way, there were no ejection seats in the early models of the airplane. Ejection seats would add thirty pounds above a regular seat, so to save precious weight, the CIA decided to dispense with them altogether.

U-2 pilots would be trained to fly 9-hour-and-40-minute missions, flying round-trip on deep-penetration flights over the Soviet Union. The pilot needed an iron butt for ten-hour flights. "I ran out of ass before I ran out of gas," some U-2 drivers would later complain — and who could blame them? A pilot was jammed inside a cockpit smaller than the front seat of a VW Beetle, laced into a bulky partial-pressure suit, his head encased in a heavy helmet, hooked to an oxygen breathing tube, a urine tube, and fighting off muscle cramps, hunger, sleepiness, and fatigue. If the cabin pressure and oxygen supply cut off, a pilot's blood would boil off in seconds at more than thirteen miles above sea level.

The U-2 was a stern taskmaster, unforgiving of pilot error or lack of concentration. No U-2 pilot, no matter how tired, would risk a few winks and leave the driving to his autopilot. The airplane demanded extraordinary pilot vigilance from the moment of takeoff. It was designed for an immediate steep climb, but it was critical to keep the wings level because they stored a very heavy fuel load and as the U-2 rose in the sky the fuel expanded under diminishing air pressure. One wing would sometimes feed the fuel into the engine more quickly than the other and that upset the airplane's delicate balances. To regain this balance the pilot had to activate pumps that moved fuel from one wing to the other. Another very tricky aspect of flying this particular machine was maintaining carefully controlled airspeed. A pilot could fly up to 220 knots during a climb with a special gust control turned on that stiffened the wings and allowed it to hit wind gusts of up to fifty knots. But he also had to guard against climbing too slowly, that is, below 98 knots, or the airplane would stall and fall out of the sky. Above 102 knots the airplane experienced dangerous Mach or speed buffeting. So the slowest it could safely go was right next to the fastest it could go as it climbed steeply to above sixty-five thousand feet. And the shuddering felt the same whether it was the result of going too fast or too slow, so a pilot had to keep totally alert while making corrections. A mistake might make the buffeting worse and shake the airplane to pieces. And to make life more interesting, our test pilots reported that sometimes during a turn the inside wing would be shaking in stall buffet while the outside wing was shaking even more violently in Mach buffet.

Once the pilot reached seventy thousand feet he tried to maintain 400 knots true airspeed, about as fast as a commercial jetliner, and keep the engine from overheating and operating at maximum efficiency.

At altitude the pilot flew nose high and wings level, so for him to be able to see down we installed a cockpit device known as a drift sight — basically an upside-down periscope that had four levels of magnification and could be swiveled in a 360-degree arc. The pilots also had to plot their navigation by sextant, plotting precise routes while maintaining total radio silence and photographing particular targets with the pinpoint accuracy of a bombardier. A screwup could mean death by ground fire or fighter attack — and a guaranteed international crisis.

The airplane and the missions were much too demanding to trust to any but the best pilots available. The CIA found that out in the late fall of 1955, when they made a totally off-the-wall decision to try to recruit foreign pilots to fly this top-secret program. The rationale was that it would be less embarrassing if, say, a Turkish national was shot down over

Russia than an American. Our government could plausibly deny any involvement. The president had cut out the Air Force from the U-2 program on the basis that the CIA was better at keeping secret a very classified program and that if a plane should be shot down it was not as provocative somehow with a civilian pilot at the controls as with an Air Force fighter pilot. Much to the chagrin of the Air Force and of several high-level CIA officials, the White House ordered the CIA to recruit pilots from NATO countries who could pass themselves off as pilots for an international high-altitude weather survey program, which was the cover story for the U-2 operation. So seven foreign pilots arrived in the late fall of 1955 and began training under the tutelage of Colonel Bill Yancey, of the Strategic Air Command, and a small crew of top-notch blue-suiter flight instructors, who had been thoroughly checked out on the U-2 by our own test pilot corps. But from the first day the undertaking appeared hopeless. The pilots lacked experience to fly such a demanding airplane as the U-2, and several of them freaked out, realizing that they would be forced to land on two tandem wheels. In less than two weeks, they were sent packing, and Kelly noted with a sigh of relief in his journal: "It's been decided to use only American pilots from now on, thank God."

Before the year ended, General Curtis LeMay, the tough, cigar-chomping commander of the Strategic Air Command, got into the U-2 act by insisting that SAC recruit the pilots for the U-2 program. LeMay was furious that his own organization was not running the program operationally and thought that Eisenhower had lost his senses by allowing the CIA to start up its own air force. He raised so much hell with Air Force Secretary Harold Talbott that he was finally cut into the deal around the edges by being tasked to hire and train the pilots from within SAC, with the additional promise that U-2s would be made available to the blue-suiters at some future time. In those days the Strategic Air Command had its own fighter wings that were used to escort its bomber force into combat. The SAC fighter pilots selected would have to resign their Air Force commissions and come to work for Lockheed as contract employees under assumed names. We would put them on our payroll and so integrate them into the company that, at the end of the line, even the KGB might have a tough time tracing any of those pilots back to the military. The spooks called this kind of total identity change "sheep dipping." This was about as close as the government and private enterprise were likely to get as teammates in top-secret espionage.

The Skunk Works would also be reimbursed by special government funds for the salaries and use of its mechanics and maintenance people who would service the U-2s at the secret overseas bases for the duration of the overflights. The agency insisted on using our mechanics over the usual Air Force crews simply because we held the monopoly on knowledge and experience on the workings of the U-2, and on these critical missions over Russia there was no margin for any mechanical failure. We needed perfectly functioning airplanes from takeoff to landing. No pancake landings on a Russian beet field, thank you.

God knows, the Skunk Works had gone out of its way to earn the agency's trust. We had even kept the production line going by putting up our own money when Congress was late appropriating money to the CIA's secret Contingency Reserve Fund. Eventually, more than $54 million was allocated for the U-2 program. Out of pure patriotism Kelly defied one of his own strictly held commandments — number 11 to be exact — which insisted that a customer's funding must be timely. We were sticklers for delivering prompt monthly progress reports to customers and keeping a close accounting of our costs. Kelly required incremental customer payouts to keep us from having to carry the government with our own bank loans. But because of the national security urgency, Kelly obtained a $3 million bank loan to cover our U-2 production costs, at a time when interest rates were only about 5 percent. Still, it was a good example of a defense contractor bailing out his government. And at the end of the line we were actually able to refund about 15 percent of the total U-2 production cost back to the CIA and in the bargain build five extra airplanes from spare fuselages and parts we didn't need because both the Skunk Works and the U-2 had functioned so beautifully. This was probably the only instance of a cost *underrun* in the history of the military-industrial complex.

The first group of six U-2 pilots recruited from the SAC fighter squadrons showed up at the Skunk Works in the fall of 1955 wearing civilian clothes and carrying phony IDs. They spent three days getting a thorough briefing on the airplane before flying off to the secret base to begin training with our test pilots. I remember talking to one of them, a nice, dark-haired fellow with a soft West Virginia accent who asked me a few technical questions about the air intakes. I would instantly recognize him four years later when his face was plastered on the front page of every newspaper in the world as Francis Gary Powers.

I learned that those pilots were being paid forty grand annually with an additional thousand a month bonus once they became active overseas. The forty grand would be held for them by our payroll department and they'd collect it only after they were mustered out. Which meant they had to survive in order to collect their just rewards.

Those pilots disappeared off my screen on the morning when they flew off to the base in a CIA-operated C-47 that had all its passenger windows blacked out. But since Skunk Works mechanics and ground crews were used exclusively to maintain the airplanes overseas, and several of my colleagues were forced to make periodic quick trips to add some new device or make a fix, we were able to keep up with the U-2 operations in fits and starts. The first contingent became operational, setting up at a base in Wiesbaden, West Germany, only ten months after the first test flight and less than eighteen months since the plane was first designed. Dick Bissell had personally obtained permission from then Chancellor Konrad Adenauer to use German soil for this secret spy operation. Simultaneously, in early June 1956, the National Advisory Committee for Aeronautics, forerunner of the NASA space program, announced in Washington the beginning of a new high-altitude weather research program using a new Lockheed U-2 airplane that was expected to fly above ten miles high. The announcement was a fraud, claiming that the new U-2 would be charting weather patterns in advance of tomorrow's jet transports. Our U-2 detachment called itself "The First Weather Reconnaisance Squadron (Provisional)." They were strange weather birds — hidden away in a remote corner of the Wiesbaden air base, guarded by CIA agents carrying submachine guns. And by the time these guys were setting up operations in Germany, with four U-2s and six pilots, we at the Skunk Works were building ten more airplanes that would supply three operational detachments: Detachment A in Germany, Detachment B in Turkey, and Detachment C in Japan.

Once that first detachment was deployed, the secrecy lid clamped shut. All of us inside the Skunk Works felt in our bones that the overflights of Russia were imminent, but only Kelly was plugged in with the CIA; he would disappear for several days and we all speculated that he was either on the scene in Germany (which was untrue) or being briefed in Washington by Mr. B (which was true) and actually shown the photos taken from the first flights (also true).

The first Russian overflight occurred on July 4, 1956. A CIA pilot named Harvey Stockman flew over northern Poland into Belorussia and over Minsk, then turned left and headed to Leningrad. He was tracked on radar all the way, and dozens of Soviet interceptors tried in vain to reach him, but he made it back safely into Germany having flown for nearly nine hours. When I came to work after that holiday weekend, Kelly sent for several of us in the analytical section and briefed us in a somewhat limited fashion. "Well, boys, Ike got his first picture postcard. The first take is being processed right now. But goddam it, we were spotted almost as soon as we took off. I think we've badly underestimated their radar capabilities. We could tell from overhearing their ground chatter that they were way off in estimating our altitude, but we always figured they wouldn't even see us at sixty-five thousand feet. And you know why? Because we gave them lend-lease early-warning radar during World War II and presumed that, like us, they wouldn't do anything to improve it. Obviously they have. I want you guys to brainstorm what we can do to make us less visible or help us go even higher."

The Soviets were launching half their damned air force to try to stop these flights, and the president was upset at how easily they were tracking the U-2. "Mr. B is trying to bunch these flights before Ike gets cold feet or the Russians get lucky," Kelly sighed. "The president has given us ten good weather days for these missions. After that, who knows?"

Other Voices
Marty Knutson

I was the first pilot selected to fly in the U-2 program and made the third flight over the Soviet Union on the morning of July 8, 1956. I was a twenty-six-year-old with a thousand hours of fighter time, who had almost died of disappointment the first time I saw the U-2. I looked in the cockpit and saw that the damn thing had a yoke, or steering wheel. The last straw. Either you flew with a stick like a self-respecting fighter jock or you were a crappy bomber driver — a goddam disgrace — who steered with a yoke, like a damned truck driver at the steering wheel of a big rig.

I wound up flying that U-2 for the CIA for the next twenty-nine years. It was a bitch to land and easy to stall out, but I fell in love. I was just crazy enough to enjoy the danger.

Now here I was flying over *Russia* in a fragile little airplane with a wingspan as long as the damned Brooklyn Bridge — and below I could see three hundred miles in every direction. This was enemy territory, big time. In those days especially, I had a very basic attitude about the Soviet Union — man, it was an evil empire, a forbidding, alien place and I sure as hell didn't want to crash-land in the middle of it. I had to pinch myself that I was actually flying over the Soviet Union.

I began the day by eating a high-protein breakfast, steak and eggs, then put on the bulky pressure suit and the heavy helmet and had to lie down in a contour chair for two hours before taking off and breathe pure oxygen. The object was to purge the nitrogen out of my system to avoid getting the bends if I had to come down quick from altitude.

I knew from being briefed by the two other guys who flew these missions ahead of me to expect a lot of Soviet air activity. Those bastards tracked me from the minute I took off, which was an unpleasant surprise. We thought we would be invisible to their radar at such heights. No dice. Through my drift sight I saw fifteen Russian MiGs following me from about fifteen thousand feet below. The day before, Carmen Vito had followed the railroad tracks right into Moscow and actually saw two MiGs collide and crash while attempting to climb to his altitude.

Vito had a *close* call. The ground crew had put his poison cyanide pill in the wrong pocket. We were issued the pill in case of capture and torture and all that good stuff, but given the option whether to use it or not. But Carmen didn't know the cyanide was in the right breast pocket of his coveralls when he dropped in a fistful of lemon-flavored cough drops. The cyanide pill was supposed to be in an inside pocket. Vito felt his throat go dry as he approached Moscow for the first time —

who could blame him? So he fished in his pocket for a cough drop and grabbed the cyanide pill instead and popped it into his mouth. He started to suck on it. Luckily he realized his mistake in a split second and spit it out in horror before it could take effect. Had he bit down he would have died instantly and crashed right into Red Square. Just imagine the international uproar!

I kept my cyanide pill in an inside pocket and prayed that I would not have an engine flameout. A flameout meant I had a pack of goddam problems on my hands that might well land me in a Russian morgue or in some goddam gulag.

I was all pumped up — like flying combat in Korea. Nothing in the cockpit was automated back then. We had to fly a precise line at seventy thousand feet, looking through the drift sight and using maps. I'd compare what I was seeing through the sight to what the map showed. Pretty damned primitive, like 1930s flying, by the seat of the pants. But we all grew very skilled at it.

I flew over Leningrad and it blew my mind because Leningrad was my target as a SAC pilot and I spent two years training with maps and films, and here I was, coming in from the same direction as in the SAC battle plan, looking down on it through my sights. Only this time I was lining up for photos, not a bomb drop. It was a crystal-clear day and about twenty minutes out of Leningrad I hit pay dirt. This was exactly what the president of the United States was waiting to see. I flew right over a bomber base called Engels Airfield and there, lined up and waiting for my cameras, were thirty Bison bombers. This would prove the worst, I thought. Because the powers that be back in Washington feared that we were facing a huge bomber gap. I proved the gap — or so I thought. As it turned out, my pictures were rushed by Allen Dulles to the Oval Office. For several weeks there was real consternation, but then the results of other flights began coming in and my thirty Bisons were the only ones spotted in that whole massive goddam country, so our people began to relax a little and we turned our attention to their missile production.

I flew hundreds of missions for the agency after that, but that moment over Engels Airfield I considered the most important of any ride I took. I was overflying the most secretive society on the face of the earth, about whom we knew little, and here arrayed below with no place to hide from my lens was a big chunk of their airpower. I remember mumbling, "Holy shit," as those cameras whirred. I knew that this was an espionage coup second to none in importance and significance.

After those first flights the Russians went all out to stop the U-2. The Russian ambassador delivered a formal protest to the State Department, and the Kremlin privately threatened the Germans to either close us down or face a rocket attack on the base. KGB agents parked in big black cars just outside the fence, watching us take off and land. So we moved to a base in southern Turkey. Most of these Turkish flights monitored Soviet missile test sites on their southern border. *Sputnik* went up October 1957, the Russians putting the first object into space orbit, and they bragged about their ICBM capabilities to reach anywhere in the U.S., although they had no test launches from May 1958 to the following February. Ike was being roasted alive by the press for letting them get the jump on us. We covered Tyuratam, their missile test facility, nuclear test sites, and Kapustin Yar, their operational center for ABMs.

I flew on the eastern side of the Urals to observe their missile test launches. The CIA had spies on the ground who tipped us off whenever there would be a missile test. We usually had one day's notice to get ready and needed the president's approval to monitor the shot. By the fall of 1959, they were test-firing one missile a week. I made one or two of those observation flights and they were truly spectacular. I flew in the dead of night over some of the most remote terrain in the world. No lights down there. On a moonless night it was like flying through an ocean of ink. I flew with a big camera perched on my lap. The camera was hand-held and had special film that could determine from the flame shooting from the rocket's nozzle what kind of fuel they were using and even how they were making their rockets. The U-2 also had special sniffers, installed on the outside fuselage, that would pick up chemical traces in the air after the firing for analysis back in Washington. Suddenly, the sky lit up and that big rocket roared off the pad. I snapped away, taking pictures of that plume for a matter of seconds before it disappeared into space. The Russians never even knew I was up there.

But the most exciting mission I ever flew was out of a small landing field at Peshawar in Pakistan, where we had a support unit set up in late 1958. The flight was so long range that there was no way for me to get back to the base. My main target was in Kazakhstan, a radar and missile test center, then on to a nuclear test site near Semipalatinsk and finally an overflight of a main ICBM launch test facility. By then I would have stretched the airplane's range to the limit and would be nearly out of gas. The plan called for me to glide over the Urals to save fuel and land at a tiny World War II airstrip near Zahedan, in Iran, right in the triangle where Afghanistan, India, and Pakistan converge. The agency would send in a C-130 with agents armed with grenades and tommy guns to secure the base from mountain bandits who controlled that territory. If I made it across the border and saw a cloud of black smoke, it meant that the field was being attacked by the bandits. If that happened, I was supposed to eject and bail out. I crossed the Russian border with only a hundred gallons of fuel remaining. Really getting hairy. I didn't see any smoke, so I came in and landed with less than twenty gallons left in the tank. One of the agents had a six pack of beer icing. They had an antenna set up and were supposed to send a coded message that I was safe. One of the guys came to me and said, "Our equipment is down. I know you're a ham operator, do you, by any chance, know Morse code?" I'm sitting there under a blazing sun, still in my pressure suit, sipping a beer in one hand, and with the other tapping out the dots and dashes.

About the start of 1959 we began seeing ominous activities going on inside Russia. Around their strategic bases were strange Star of David patterns. We learned quickly enough that that meant the construction of ground-to-air missile sites. We now had orders that if we saw any new Star of David patterns, we were to deviate from the flight plan and go film them. These first SAMs couldn't reach us. The optimum use of their surface controls was lower — fifty-five thousand feet — to be used against our bombers. But we figured we were flying on borrowed time. Sooner or later, one of us would get nailed. I knew for sure it would be the other guy.

They tried to stop us by trying to ram us with their fighters like a ballistic missile. They stripped down some of their MiG-21s and flew straight up at top speed, arcing up to sixty-eight thousand feet before flaming out and falling back toward earth. Presumably they got a relight down around thirty-five thousand feet. I'm sure they lost some airplanes and pilots playing kamikaze missile. It was crazy, but it showed how angry and desperate they were becoming.

By the winter of 1960, we were getting intelligence briefings warning us about improvements in Soviet tracking and SAM capability. Their new SA-2 missile was an improved version of what they previously had, capable of reaching us and equipped with a powerful warhead that could be lethal if exploded within four hundred feet of an airplane. We gave the SA-2 a wide berth whenever possible. I have to admit we were all getting plenty worried. We had long since installed ejection seats. That was one item of added weight no one in his right mind would do without.

The CIA code-named the project Rainbow. The orders came directly from the Oval Office and had the highest priority. Every one of Kelly's engineers and designers was put to work.

The object: significantly lower the U-2's radar signature or face a presidential cancellation of the entire program. These orders arrived just before the new year in 1956, after only seven or eight overflights of the Soviet Union. The Russians were using diplomatic channels to scream at us. They were too embarrassed by their own ineptitude in being able to stop the overflights to make their threats public, but that did not make these threats less ominous.

So the heat was on to find ways to reduce the airplane's radar signature from approximately that of a Fifth Avenue bus to the size of a two-door coupe. But the U-2's big tail, wings, and large inlets designed for its lift and thrust got in the way, acting like a circus spotlight on hostile radar scopes. Kelly flew to Cambridge, Massachusetts, seeking advice from a high-powered scientific group doing research on antiradar technology. He brought back with him two of their best radar experts, Dr. Frank Rogers and Ed Purcell, to help us brainstorm. I was involved analyzing various composites and paints that might be tried to absorb radar energy without adding too much weight to the airplane. The U-2 was painted a dull black to prevent it from glinting in the sun. We wanted to make it as difficult as possible to achieve human detection from the ground or from below in an interceptor trying to reach it. And we began experimenting with chromic paint that changed from different shades of blue to black at different temperatures like a chameleon. But paint added more weight than deception, as did the idea of painting the U-2 with polka dots to break up the silhouette against the sky. As for fooling radar, Rogers and Purcell suggested a radical fix: stringing piano wire of various dipole lengths along the entire fuselage in the hope of scattering as much radar energy in as many frequencies as possible in every direction. The wires made the U-2 draggy and we lost seven thousand feet in altitude.

The next thing we tried was something called a Salisbury screen, a metallic grid applied to the airplane's undercarriage in the hope of deflecting incoming radar beams, but it worked only at some frequencies and altitudes and not at others.

Kelly thought it was more practical to try special iron ferrite paints that would absorb a radar ping rather than bounce it back to the sender. The paints were moderately effective but inhibited heat dissipation through the airframe's outer skin and we experienced overheating engine problems. But the paint lowered the radar cross section by one order of magnitude, so we decided to give it a try. We called these specially painted airplanes "dirty birds" and shipped the first one out for flight testing in April 1957. Our test pilot, Bob Sieker, took the U-2 up to over seventy thousand feet and suddenly radioed that he was experiencing rapid airframe heat buildup. Moments later his engine blew out and the faceplate blew off Bob's oxygen mask as his pressure suit instantly inflated. The U-2 dove straight down and crashed. It took us three days to locate the wreckage and Bob's body. An autopsy revealed that above seventy thousand feet he had suffered acute hypoxia and lost consciousness in only ten seconds. The culprit that killed him was a defective faceplate clasp that cost fifty cents.

The CIA was so desperate to buy time for these Soviet overflights that Bissell got Kelly to sequester four of our test flight engineers and have them write a bogus flight manual for a U-2 twice as heavy as ours and with a maximum altitude of only fifty thousand feet that carried only scientific weather gear in its bay. The manual included phony instrument panel photos with altered markings for speed, altitude, and load factor limits. Four copies were produced and then artificially aged with grease, coffee stains, and cigarette burns. How or if the agency got them into Soviet hands only Mr. B knew, and he never told.

Other Voices
James Cherbonneaux

In July 1957, a cargo plane brought the first so-called dirty bird to our base in Turkey. It was covered with a plastic material and had two sets of piano wire strung from either side of its nose to a set of poles sticking out of the wings. The wires were to scatter radar beams while the paint was to absorb other frequencies. But I wasn't thrilled. Part of my big paycheck was compensation for high-risk missions in a semi-experimental airplane, but I had never before risked flying an airplane wired like a guitar.

The Skunk Works engineer who flew out with the dirty bird admitted that its extra weight would cost us altitude and three thousand miles in diminished range. On July 7, 1956, I flew the dirty bird on an operational test of the whole Soviet defense net along the Black Sea, flying inside twelve miles of the coastline. The flight lasted eight hours, and I carried an array of special recording devices while deliberately trying to provoke responses from Soviet air defense along its entire southern flank. The plan was to see if they could detect our dirty bird. All in all, the coatings and wire worked well, but analysis of my recordings indicated that the bad guys were homing in on my cockpit and tailpipe, neither of which had been treated.

As it turned out, my most incredible overflight occurred only two weeks later in a mission specially cleared by President Eisenhower and involving a dirty bird. I took off in total secrecy from a field in Pakistan, flying a dirty bird for a flight deep inside Russia to photograph a missile site believed to be readying intercontinental missile tests. Because of the added "dirty" weight I could top out at only fifty-eight thousand feet. The missile site was a three-hour trip, but about seventy-five minutes into the mission I looked through my drift sight and saw a startling sight: the familiar circular graded contours that I had seen marking our own nuclear test site at Yucca Flats. My heart jumped! *Could it really be?* We had no idea that this test site even existed. I brought my drift sight up to its maximum four power magnification and focused on a large tower. I felt a chilling terror. That tower held a large object at the top, and there were signs of activity around a huge blockhouse about two miles away. And then a paralyzing thought slammed me: what if those bastards were getting set to let that nuclear weapon blow just as I was directly overhead? And in fact, that crazy thought took hold and I began to sweat and hyperventilate in panic. "Wait, goddam it. Wait, will you! Let me pass and then light your fire." I was shouting into my faceplate.

I carried a three-camera system, one pointing straight down and the other two out at forty-five-degree angles so that each picture would overlap and provide a stereographic photo interpretation. I threw a switch and the cameras began to whir in sequence. It seemed to take an eternity for my airplane to cross directly over that tower. My heart was pounding in my throat. I just knew I was going to be evaporated in the next seconds.

Five minutes later I was clear of the nuclear test site, laughing at myself for being so chicken. Three hours later I was over the city of Omsk, in central Russia, photographing a military-industrial complex of interest to SAC as a potential target; then I turned east to head for the missile test site — my principal target for the mission. I photographed the site, which bore evidence of a very recent test firing, then headed back to Pakistan.

I looked through the drift sight at the vastness of central Russia, a vastness almost unimaginable that made me feel achingly lost and alone. There were no telltale contrails of MiGs trying to get me, so I had to conclude that the Skunk Works had worked its magic and kept me hidden from the bad guy's radar. If true, that meant that no one on earth knew where I was at that particular moment because I was also out of range of U.S. listening posts. Suddenly I became alert. My engine started making rough noises, but I knew from long experience that the roughness of an engine is in direct proportion to how far a pilot still has to go before he makes it safely across a hostile border.

Twenty minutes later I saw the shimmering snowy peak of the awesome mountain called K-2 illuminated against a dark blue sky. K-2 was my beacon back to Pakistan and a hot shower and sleep, and I calculated it was an hour away, about 450 miles, before I crossed the border. I was now eight hours plus into the mission and I became aware of the need to urinate. I cursed myself for forgetting to avoid any liquids the night before the flight, something I tried always to do because I was never able to pee out of my pressure suit. To do so, I had to unwork three layers and then pee uphill into a nozzle arrangement. I just couldn't. But I began to ache. Real pain. Like knife thrusts down there. I was exhausted and in agony. I tried to pee into the nozzle, I tried to wet my pants, but I was having spasms and nothing came out. It was so bad I could barely focus my eyes, and K-2's magnificent granite towers slipped by directly below and I barely noticed. Or cared a damn. By the time I made my landing approach the pain was searing to the point where I almost landed short of the field and crashed into a forest. I barely remember the mechanic opening my canopy at the top of his ladder and me pushing him aside like a maniac and vaulting down that ladder in a flash. Moments later, not heeding privacy, I set a new world record on that tarmac.

I expected to be received as a hero for having uncovered a Soviet nuclear test site with a bomb in the tower. Instead the team who debriefed me scoffed incredulously. "There is no atomic test facility in that part of central Russia," one debriefer told me. But they passed on my observation by special wire to Washington, while another team began processing my film. CIA headquarters responded by coded cable in less than an hour. Their communication was stern, halfway between a personal rebuke and an official reprimand. My credibility was zero back home. But the next day over lunch, John Parangosky, the senior CIA agent in charge of our operation, took me aside with a sheepish grin. "Apologies, Jim. Collateral intelligence sources just reported that a nuclear bomb was detonated from that tower less than two hours after you flew over it."

During the final winter of the U-2 overflights of Russia, Kelly Johnson came back from a visit to CIA headquarters looking profoundly gloomy. He couldn't believe how easily the Russians were tracking our overflights and knew it was only a matter of time before their ground-to-air missile defenses caught up with their prowess in radar development and blew us out of the sky. "Putting fixes on this airplane won't do any good. We need a fresh piece of paper," Kelly told a group of us. His mind was already churning, thinking about the U-2's successor that could survive flying above Moscow. He had asked our ace mathematician Bill Schroeder to predict how long it would take the Soviets to bring down a U-2 with their latest missile system. Schroeder gave the U-2 less than a year.

One of our engineers came back from a quick-fix visit to the secret U-2 base in Turkey to say that the morale among the pilots was sagging. The guys were worried about new SA-2 missile sites under construction around the Soviet Union's main target areas. The president was very aware of the growing dangers and had cut back sharply on authorizing U-2 missions. And the trip our engineer made to Turkey indicated the growing concerns about pilot safety: he was on hand to supervise a new "black box" installed into the U-2's tail section to electronically counter incoming radar beams and scatter them away. In the jargon of the trade, the box was called an ECM — electronic counter-measure — and would hopefully prevent a missile fired at a U-2 from locking on.

The word Kelly received from Dick Bissell was that the intelligence community was pushing hard for at least one more overflight over Tyuratam, the big Russian missile test center deep in the Urals, since a recent flight had revealed significant advances toward development of their first operational intercontinental missile. Eisenhower was ready to approve the follow-up flight, but the State Department heard about it and Secretary of State Christian Herter was strongly opposed. Herter had replaced John Foster Dulles, who had died of cancer earlier in the year, and was worried that any overflight might upset the delicate planning that had revolved around a summit in Paris between Ike and Khrushchev scheduled to start on May 14.

Bissell told Kelly that Allen Dulles had wrested one final flight out of the president, provided it took place two weeks before the Paris summit. The target date was May 1, 1960, the Soviet May Day, akin to our Fourth of July. We hoped to catch them with their defenses down, with only skeleton crews at work.

As it turned out, our black box and the route of the mission finally selected would seal the fate of that tragic last flight. Ike had signed off on two mission options and left the final decision to the CIA. The choices were missions code-named Time Step, which would overfly certain key nuclear and missile test sites, and Grand Slam, a marathon nine-hour mission from Pakistan clear across Russia to land at a base in Bodo, Norway. The heart of Grand Slam was overflying Tyuratam, then heading south to photograph the huge military-industrial complexes at Sverdlovsk and Plesetsk. All were heavily defended.

The two plans were sent for review to the Air Force chief of staff, General Nathan Twining, who quickly spotted a flaw in the Grand Slam mission and called Allen Dulles to personally urge changes. Twining had noticed that the proposed mission repeated the exact route into Sverdlovsk from the south used less than a month earlier by U-2 pilot Marty Knutson. "Allen, if you come in that way again, they'll know exactly where you are heading and will just be lying in wait. You'll get nailed."

Dulles obviously didn't agree. He personally chose Grand Slam with no changes.

Because the mission would be so demanding and long, covering 3,700 miles from Pakistan to Norway, the agency chose its most experienced pilot and the best navigator of the group — thirty-four-year-old Francis Gary Powers. The pilot had twenty-seven U-2 missions logged, including several marathon-length flights across the eastern Mediterranean in 1956 to gather intelligence on the movements of British and French warships participating with Israel in attacking Egypt during the Suez crisis.

Powers took off at dawn from Peshawar, Pakistan, on Sunday, May 1, 1960. Flying across the Soviet border for the first time from Pakistan was another way to catch the Russians napping. And for the first three hours into the flight the plan worked perfectly. He flew over Tyuratam without difficulty then changed course and headed south toward Sverdlovsk, on the same flight plan as Knutson's only weeks earlier. As he approached the Sverdlovsk complex, Powers was suddenly blinded by a brilliant orange flash and felt an explosion from behind. His right wing dropped and he began pitching down. His instincts told him his tail had been hit as the airplane began a steep nosedive. In horror he saw his wings rip off. His pressure suit inflated, squeezing him in a viselike grip, and his faceplate began to frost. He glanced at the altimeter, saw he was at thirty-four thousand feet and falling fast, and almost panicked realizing he was pinned by the centrifugal force up against the instrument panel. If he hit the ejection lever, he'd be blasted out of the cabin while leaving both his pinned legs behind. He struggled to push back in his seat and manually open the canopy. He unhitched his safety harness, and as the wingless fuselage spun upside down, Francis Gary Powers fell free.

As his chute opened, Powers was startled to see another chute opening in the distance. Whatever hit him had also hit a

Soviet pilot as well. He landed hard in a farmer's field. Several villagers came running to him. They weren't unfriendly and had no idea he was an American because he was too stunned to even say a word while they conversed among themselves excitedly. They finally helped him to a truck and drove him off. He would later learn they were driving him to the local airport, assuming he was a Russian pilot and not knowing what else to do with him. At some point, though, the truck was stopped by the militia. The police grabbed Powers and took him away.

It would later be determined that a Soviet missile battery had launched in shotgun fashion fourteen SA-2s at the approaching U-2 — an indication that they were waiting for his arrival. One missile had knocked down a Russian fighter trying to intercept Powers, and the shock waves from the exploding missiles had knocked off the U-2's tail.

Kelly received the call at home, well after midnight, and he grimly arrived at the Skunk Works that Monday morning and assembled a group of us. "We got nailed over Sverdlovsk by an SA-2. That's that. We're dead."

It was the first time in history that a ground-to-air missile had shot down an airplane, and all of us assumed, knowing how fragile the U-2 was and at the height it was probably flying when it was hit, that the pilot had been killed. The CIA immediately had NASA launch a preplanned cover story that one of its weather research planes, flying out of Turkey, had strayed off course and was missing after the pilot indicated he was having oxygen problems. Cagey Khrushchev waited for Eisenhower to arrive in Paris for the summit before announcing that the Russians had shot down a U-2 spy plane. The administration called that a "fantastic allegation." Eisenhower denied spy flights, and then on the eve of the sum-

mit Khrushchev announced that the pilot had been captured
and confessed his spy mission. The pilot was named Francis
Gary Powers.

Eisenhower was humiliated and forced to admit the U-2
spy operation, which he said was justified since Khrushchev
had recently turned down his Open Skies proposal. To mollify
the Russians and save the summit, Ike announced we would
end the flights, which he had privately done anyway. But when
Khrushchev demanded an apology, the summit collapsed and
Eisenhower went back home.

Inside the Skunk Works we were no less stunned that
Powers had survived than the CIA and the White House. The
agency was livid at Powers for not dying in the hit or taking
his own life, even though using the poison needle that had
replaced the cyanide pill in a pilot's kit was entirely optional.
But some of the more macho patriots around the Skunk
Works agreed with their opposite numbers thundering
around at the CIA that Powers was a damned traitor for not
self-destructing. And they meant it! Because he was chicken,
the president endured a terrible international humiliation.
Powers's survival also embarrassed Dulles and Bissell, who
had assured the president, presumably in good faith, that not
much would be left of a U-2 or a pilot if shot down by a missile. Powers was also faulted for not pulling a seventy-second
delayed explosive charge before bailing out that would have
destroyed the film and cameras and kept them out of the
hands of the KGB.

There was little sympathy for Powers, who was kept incom-
municado inside the notorious Lubianka prison for months
before enduring a propaganda show trial that heaped embarrassment on the agency and the administration for more than
three weeks. Powers was sentenced harshly to ten years at
hard labor and served nearly two years before being ex-
changed in February 1962, for the captured Russian master

spy Rudolph Abel, a decision that only enraged many at the CIA even more. "That's like trading Mickey Mantle for a god-dam bullpen catcher," one of the agency guys exploded when hearing the news.

Had Powers killed himself or not survived the missile hit, he would have come home a hero in a flag-draped wooden box. But coming home haggard and alive, he was greeted like a traitor and was whisked off in great secrecy to a CIA safe house in Virginia to be grilled unmercifully for days about his experiences over and inside Russia. Kelly was summoned to the debriefing to hear the part about the shoot-down and was satisfied that Powers was telling the truth.

Kelly had long ago analyzed photographs of the U-2 wreckage released by the Russians and reported to Bissell his conviction that the airplane had been hit from the rear. "It looks like they knocked off his tail." At the debriefing, Powers confirmed that fact. Kelly felt sorry for the guy and offered him a job as a U-2 flight test engineer at the Skunk Works. He gratefully accepted and worked for us for eight years, until the mid-1970s, when he went to work for a local TV station as a helicopter traffic reporter. He was killed in a helicopter crash on August 1, 1977. Ten years later the Air Force awarded the former captain a posthumous Distinguished Flying Cross, a medal well earned if sadly late in arriving.

Kelly long suspected that the electronic counter-measure black box we installed on the tail section of Powers's U-2 may have acted in an opposite way from the one we intended. The box was code-named Granger, and we provided the frequencies used to jam and confuse the enemy missile. These were the same frequencies the Russians used on their defensive radar. But it was possible that the Russians had changed these frequencies by the time we incorporated them into our missile spoofer, so that the incoming missile's seeker head was on the same frequency as the beams transmitted off our tail and acted as a homing device. A few years later a similar black box

was installed in the tails of CIA U-2s piloted by Taiwanese flying highly dangerous missions over the Chinese mainland.
One day three of four U-2s were shot down, and the sole sur-
vivor told CIA debriefers that he was amazed to be alive because he forgot to turn on his black box. To Kelly, that
clinched the case. But we'll never really know.

Other Voices
Richard Helms
(Director of the CIA from 1966 to 1973)

The U-2 overflights of the Soviet Union provided us with the
greatest intelligence breakthrough of the twentieth century.
For the first time, American policymakers had accurate, credible information on Soviet strategic assets. We could evaluate
in real time the other side's strengths and weaknesses, keep
current on their state of preparedness, their research and development, their priorities in defense spending, the state of
their infrastructure, and the disposition of their most important military units. It was as if the scales had been lifted from
our eyes and we could now see with clarity exactly what it was
we were up against. It really was as if we in the intelligence
community had cataracts removed, because previous to those
splendid U-2 missions our ability to pierce the Iron Curtain
was uncertain and the results were often murky. We were
forced to use defector information and other unreliable means
to sift for clues about what the other side was up to. Given how
little solid information actually filtered out to the West, we did
a credible job, but the U-2's cameras leapfrogged us into another dimension altogether. For example, those overflights
eliminated almost entirely the ability of the Kremlin ever to
launch a surprise preemptive strike against the West. There
was no way they could secretly prepare for war without our
cameras revealing the size and scope of those activities.

Building the U-2 was absolutely the smartest decision ever

164 SKUNK WORKS

made by the CIA. It was the greatest bargain and the greatest triumph of the cold war. And that airplane is still flying and is still tremendously effective. In my opinion, the national security demands that we keep supplying new generations of surveillance aircraft to our policymakers. There is no way to replace the vital data provided by piloted airplanes. Satellites lack the flexibility and the immediacy that only a spy plane like the U-2 can provide. No president or intelligence agency should have to operate with only one eye in such an uncertain and dangerous world.

Richard Bissell

I have no doubt that the U-2 overflights of the Soviet Union made up the most important intelligence-gathering operation ever launched by the West. Until those flights, our side had to be content with some ingenious analysis on our part about their nuclear program, for instance, that later U-2 overflights would confirm as being remarkably correct. We were much less correct about their missile development because we had assumed — quite incorrectly — that they would continue to develop liquid-fuel missiles, while very secretly they dropped that concept and embarked on more sophisticated, solid-state missiles. That caught us by surprise and generated the so-called missile gap.

There was also a profound worry about the size of their long-range bomber fleet. President Eisenhower told Allen Dulles that obtaining a hard count of their bombers was the urgent priority of the intelligence community. And by the time Allen chose me to head the U-2 project, the president told me that he regarded hard intelligence on Russian bombers as the number one item on his national security agenda. He told me that the minute I flashed the signal to him that Kelly Johnson was ready to deliver that airplane, he was ready to give me permission to start those flights.

I told the president that we would probably have two years

before the Russians would find a way to bring us down. As it turned out, we had a fruitful four years.

The first flights I decided to bunch. My reasoning was that the first would be the safest, catching them by surprise, so we'd overfly all the highest-priority targets. The first flight was to be over Leningrad, picking up important missile test sites and air bases along the route, then fly the length of the Baltic coast. I stopped by Allen Dulles's office and told him, "Well, we have an Oval Office green light and we're off and running." When I told him the flight plan, he turned deathly pale. A few hours later I was able to inform him that all went well. The next day we scheduled two separate flights, one into the Ukraine and the other well north of that. We were looking for military airfields — our primary target. Only in later months did the location of hardened missile silos take precedence.

It took us four days to get our hands on the photographs from that first mission. I remember vividly standing around a long table with Dulles next to me, both of us chuckling with amazement at the clarity of those incredible black-and-white photos. From seventy thousand feet you could not only count the airplanes lined up at ramps, but tell what they were without a magnifying glass. We were astounded. We had finally pried open the oyster shell of Russian secrecy and discovered a giant pearl. Allen rushed with the first samples over to the Oval Office. He told me that Eisenhower was so excited he spread out the entire batch on the floor and he and Allen viewed the photos like two kids running a model train.

We never knew what we'd find from one mission to the next. Every airfield discovered increased our knowledge dramatically. On one flight a pilot saw a railroad track in the middle of nowhere and followed it and brought back stunning pictures of a Soviet missile launcher at a site we never knew existed. Many other photos were confirmations of locations of important military bases that we had received from our spy network on the inside. We would get a tip about a new plant somewhere, but

the informant was uncertain about whether they were manufacturing tanks or missiles, so we would schedule a look. Our
first missions out of Pakistan were staged so that we could
overfly central Siberia and observe the Trans-Siberian railroad,
mainly because we had very sketchy inferential information
that atomic facilities were being built there. We brought back
very revealing photos indicating that a nuclear test facility was
nearly completed on the site.

After only four or five flights our analysts were able to make
much firmer estimates of the Soviet bomber strength by types.
We had a count on how many planes of each type were photographed sitting out on their ramps. Of course, it was not water-
tight because airplanes seldom stayed put at one base or
another, and it was hard to tell if we were counting the same
airplanes seen at base A that now appeared at base B, or if
these were additional ones. But the accumulated weight of evidence from these flights caused the president of the United
States to draw in a deep breath, smile, and relax a bit. I was
able to assure him that the so-called bomber gap seemed to be
nonexistent.

By six months into the overflights we turned our attention to
their missile development. We found big research and development bases at the head of the Caspian Sea and just east of the
Volga, and saw hard evidence of a number of experimental
launches that had taken place there. We found a big radar installation at Sari Sagan, between Turkistan and Siberia, and
also a down-range site under construction near there, so we
began to monitor this particular section very closely.

Our estimates of their SAM missiles was that they could
reach the altitude of the U-2 but that their surface controls
were effective only up to fifty-five thousand feet and any higher
than that they couldn't control a missile and bring it home for
a kill against our spy plane.

But the very unpleasant surprise was the ease with which
they tracked every single one of our flights — almost from

takeoff. Yet, until the Powers flight, they had never come close to hitting us. On one night flight out of Turkey they had actually scrambled fifty-seven fighters against one U-2. And on many occasions they were flying squadrons fifteen thousand feet underneath the U-2, trying to block the view. Kelly Johnson called that "aluminum clouds."

After the first few flights they tracked, they could infer the U-2's range, speed, altitude, and radar cross section, so they knew all the important essentials about the airplane which we cloaked under the deepest secrecy.

Ironically, the two governments, in their abiding hostility, were collaborating to keep these flights secret from the public. Because if they were ever revealed, the Russians would have to present us with an ultimatum and admit that they were impotent in stopping these flights over their territory. It must have been terribly upsetting inside the Kremlin knowing that the enemy could overfly with impunity. So I was constantly pressing Eisenhower for more flights and he was constantly resisting me. I had to go to the mat on nearly every authorization because he was following the advice of the other Dulles brother, John Foster, our secretary of state, who was wringing his hands over the spy flights right from the beginning.

We flew fewer than thirty missions over those four years, but each of them was a remarkable success. We accumulated about one million two hundred thousand feet of film — a strip almost two hundred and fifty miles long, that covered more than a million square miles of the Soviet Union. The flights provided vital data on the Soviet atomic energy program, their development of fissionable materials, their weapons development and testing, and the location and size of their nuclear stockpile. It also gave us precise information on the location of their air defense systems, air bases, and missile sites; submarine pens and naval installations; their order of battle, operational techniques, and transportation and communications networks. By the Pentagon's own estimate, 90 percent of all hard data on Soviet military development came directly from the cameras on board the U-2.

As early as three years before Powers was shot down, I flew out to Burbank with my deputy, Colonel Jack Gibbs, to meet with Kelly about the future. We estimated that the U-2 was operating on borrowed time after the two-year mark. I said to Kelly, "We've got to begin now to design a successor." He told me he had already begun thinking about a liquid hydrogen–powered airplane and was looking at ways to make his own liquid hydrogen fuel and build his own tank farm. A hydrogen-powered airplane was certainly ambitious, and in those days Kelly seemed entirely capable of moving the world.

8

BLOWING UP BURBANK

IN THE EARLY WINTER of 1956, Kelly sent for me, and I walked down the hall to his office with my heart in my throat. I feared he was sending me back to the main plant with a handshake and goodbye. "Close the door," he said.

I sat down opposite his desk like a condemned man sensing the verdict.

"Rich," he asked, "what do you know about cryogenics?"

I shrugged. "Not much since college chemistry days."

"I want you to read up thoroughly on all those exotic fuels, especially on liquid hydrogen, and then get back to me and we'll talk some more. Keep your damn mouth shut about this. Tell no one."

When Kelly had tapped me for this assignment, the first thing I had done was to check the reference to liquid hydrogen in my copy of Mark's *Mechanical Engineering Handbook*, the engineer's bible, which told me what I already thought I knew — liquid hydrogen had no real practical application because it was so dangerous to store and handle. It was a mere laboratory curiosity. I read that definition to Kelly Johnson and told him that I happened to agree. Kelly's face reddened — a storm was rising. "Goddam it, Rich, I don't care what in hell that book says or what you happen to think. Liquid hydrogen is the same as steam. What is steam? Condensed water. Hydrogen plus oxygen produces water. That's all that liquid hydrogen really is. Now, get out there and do the job for me."

Over the next few weeks I was living a boyhood fantasy and traveling around the country pretending to be a secret agent, using my Skunk Works alias of "Ben Dover," in the best traditions of trench-coated operatives. Kelly had warned me not to reveal that I worked at the Skunk Works to anyone I visited. I pretended to be a self-employed thermodynamicist trying to learn as much as I could about liquid hydrogen for an investment group studying the possibilities of a hydrogen airplane engine. I was consulting with hydrogen experts around the country to find out how we could make our own liquid hydrogen safely and cheaply in large batches to fuel Kelly's latest dream. He was thinking about building a liquid hydrogen–powered spy plane as the successor to the U-2, giving us twenty times the thrust and power of a conventionally powered airplane. We'd be practically a space vehicle, whisking above 100,000 feet across the sky at more than twice the speed of sound. The entire Russian defense system would fall into a state of catatonic disbelief, mistaking us for a streaking comet. "I want answers, not excuses about why we can't do this," Kelly told me and shoved me out the door.

Which is why I showed up as Mr. Ben Dover at Boulder, Colorado, where the U.S. Bureau of Standards maintained a cryogenic laboratory under the direction of Dr. Russell Scott, recognized as the world's expert on handling and storing liquid hydrogen. When I told him I wanted to learn how to handle liquid hydrogen in large amounts — like maybe running my own tank farm — the blood drained from Scott's face. "Mr. Dover," he said, "this stuff is *volatile*. One tank car could blow up an entire shopping mall. Do you have any notion of the risks?"

My fact finding took me into the dark and gloomy basement of the chemistry building at Berkeley, where Nobel Laureate William Giauque held forth from a reinforced basement bunker doing his prize-winning experiments on low temperature research. I couldn't help noticing some holes punched in the walls, courtesy of errant handling of small teacup amounts of liquid hydrogen by student lab assistants. "Handle with extreme care, Mr. Dover," Professor Giauque warned me. "That's why they keep me stashed away in this dungeon." When I told him that I wanted to learn how to make liquid hydrogen and store it in the hundreds of gallons, the professor shook his head solemnly. "With all due respect, sir, I think you've got a screw loose."

I wanted to protest: "Not me, Prof, but that lunatic I work for."

Actually, the idea of using hydrogen as a propellant had been kicking around since the end of World War II, primarily to fuel rocket engines, simply because its volatility created tremendous thrust. But Kelly wasn't thinking of a rocket engine; he wanted a conventional jet engine fueled by liquid hydrogen that could cruise for hours above Mach 2. Rocket engines were indeed like comets — blazing into the sky for a minute or two before extinguishing. As a hard-nosed businessman Kelly was not about to commit to building such an airplane unless he felt assured that we could produce sufficient supplies of fuel and learn how to handle it safely. So when I returned to Burbank armed with blueprints and technical manuals, Kelly smiled upon me benevolently and told me I was in charge of building our own hydrogen liquefaction plant.

He sent me to a remote corner in the Lockheed complex, which had been during World War II a communal air raid shelter for hundreds of workers in the nearby B-17 bomber factory. Since then, it had been used to store bombs and bullets for our flight test division. It had eight-foot-thick walls and underground bunkers and was, I noticed, as far away as he could get from us in case something went wrong and we blew up. "Here's what I want," he told me. "I want to show the blue-suiters that working with liquid hydrogen is not so risky once you attain experience handling it. I want to prove that a damned Air Force airman can handle it as well as a Ben Rich."

The reason he was mentioning the Air Force was that the CIA had quickly rejected the idea of building a hydrogen airplane as the successor to the U-2. Bissell had had his own bean counters estimate the development costs, which were in the $100 million range — too costly for the agency's secret contingency funding, which bypassed the usual congressional appropriations committees. Even the CIA would find $100 million hard to hide. The Air Force, smarting for having to play a passive role in the U-2 Russian overflight operation, was receptive. The hydrogen airplane would put the blue-suiters in the driver's seat for the next round of spy flights. Kelly had a promising preliminary discussion about the concept with the top Air Force brass involved with planning and development. They were eager to work with Kelly Johnson on just about anything involving Soviet overflight operations and decided to fund a feasibility study and begin the process of selecting a manufacturer to build a hydrogen-fueled engine.

I was a key player in the feasibility study. I had to prove that the fuel was safe and practical to produce in large batches. Kelly wanted me to try to create controlled explosions and fires in order to learn what we were up against. I requested Dave Robertson to help me. Davey was one of the shrewdest, most instinctive engineers I had ever known, with the right flair for these wild experiments.

Thank God my wife, Faye, had no inkling of how I was earning my paycheck in the late autumn of 1959. I remember huddling behind cement barricades with Robertson, trying to create a "controlled" explosion by rupturing tanks filled with liquid hydrogen under pressure. Nothing happened. The hydrogen just escaped into the atmosphere. So we set a charge and ignited it. Because of its low density the fireball quickly dissipated. The biggest bang, which knocked us four feet backwards, came when we mixed liquid oxygen with an equal amount of liquid hydrogen. The shock wave thudded against a huge hangar under construction about five hundred yards away and nearly knocked four workers off the scaffolding, while Davey and I huddled out of sight behind the cement wall, giggling like schoolboys.

One of our colleagues named our walled-in compound Fort Robertson because the guy and the place seemed perfectly mated, and the name stuck. We got Dr. Scott of the Bureau of Standards cleared to work with us as an adviser. The Fort Robertson complex was located less than a thousand yards from the Municipal Airport's in-bound runway. And the first time Dr. Scott paid us a visit and saw the three tanks of liquid hydrogen holding hundreds of gallons under storage, his knees began to shake. "My God in heaven," he exclaimed, "you're gonna blow up Burbank."

Scott came up with a brilliant idea. He suggested we substitute liquid nitrogen, which was less volatile and dangerous than liquid hydrogen, in our experiments as a safer substitute to test what might happen if we used liquid hydrogen under certain conditions. We made about twelve hundred gallons of liquid nitrogen. We made a martini in a dixie cup, then dipped a popsicle stick in the liquid nitrogen and used it to stir the martini. It became a popular taste treat for those cleared to visit us.

In less than three months, working with twelve Skunk Works shop workers and mechanics, we began producing more liquid hydrogen than any other place in the country — about two hundred gallons daily. We stored it in a ten-foot-high tank capable of pumping six hundred gallons a minute. We wore special grounded shoes and couldn't carry keys or any other metallic objects that might spark. We installed a nonexplosive electrical system and used only nonsparking tools. Dave Robertson also invented a special hydrogen leak sniffer around the tanks that would immediately sound a klaxon horn warning that would send us running — probably for our lives.

Kelly was pleased with our progress. On the drawing boards was a design for the dart-shaped CL-400 that would fly at 100,000 feet at Mach 2.5 with a 3,000-mile range. The body was enormous, dwarfing any airplane on the drawing boards. On the playing field at Yankee Stadium, for example, the tail would cover home plate and the nose nudge the right-field foul pole, 296 feet away. It was more than twice the size of the B-52 bomber. And the reason the body was so gigantic was that it would carry a fuel load of liquid hydrogen weighing 162,850 pounds, making it the world's largest thermos bottle. Flying at more than twice the speed of sound, the outer shell of the body would blaze from heat friction above 350 degrees F while the inside skin would hold the frosty fuel at temperatures of minus 400 F — an 800-degree temperature differential that represented an awesomely complicated thermodynamic problem. Undaunted, Kelly promised to have a prototype ready in eighteen months.

The Air Force allocated $96 million, and we were off and running. The code name was Suntan, and it was classified *above* top secret. That was a first even for us. Only twenty-five people at the Skunk Works were cleared to work on it. And around the time that the U-2 overflights of Russia revealed that no bomber gap existed, the CIA dropped a bombshell by presenting Eisenhower with strong indications that the Russians were crashing development of a hydrogen-powered airplane of their own.

They had released from a Siberian gulag a brilliant scientist named Pyotr Kapitsa, who had been arrested by Stalin in 1946 for refusing to work on their atomic bomb development. Kapitsa was Russia's foremost expert on liquid hydrogen. He was now back in Moscow working on a top secret program. Allen Dulles and Dick Bissell agreed that it was likely the Russians were rushing development of a liquid hydrogen–propelled interceptor that could easily climb to the U-2's heights and shoot it down. The CIA came to Kelly and asked his opinion. "They might be working their tails off to get this airplane into production," he told Dulles, "but they won't have a prototype finished in less than three years or even longer."

So suddenly we found ourselves in a contest with the Russians to build the first hydrogen-powered airplane. The Air Force contracted with Pratt & Whitney to build the engines at its Florida complex, and a special hydrogen liquefaction plant was constructed to fuel the engine tests. The Massachusetts Institute of Technology was working on an inertial guidance system, and Kelly ordered two and a half miles of aluminum extrusions in advance of construction.

But six months into the project the furrows deepened on Kelly's brow. He was growing increasingly concerned that the airplane would not have adequate range to get the job done. "We've crammed the fuselage with as much fuel as it can hold," he complained at our weekly progress meeting, "and we can't extend the range by more than twenty-five hundred miles." The problem was complicated by the fact that the Air Force engineers at Wright Field had come up with wildly more optimistic figures. They predicted a range of thirty-five hundred miles, which was more than acceptable, matching the U-2's. Kelly stepped up design changes and wind tunnel testing of various wooden models but remained convinced that Wright Field's calculations were dead wrong.

Meanwhile I was having troubles of my own. Inside a hangar I built a half-scale model of the fuselage and constructed a double-walled fuel tank. I wanted to simulate supersonic flight temperatures, so I installed a wooden-framed oven over the fuselage to heat it to 350 degrees F or higher. And on a clammy spring evening in late 1959, the damned stove caught on fire only a few feet away from a storage tank containing seven hundred gallons of liquid hydrogen. We tried to put the fire out with commercial fire extinguishers but they had little effect. I sure as hell was reluctant to call the Burbank Fire Department and have them discover all that liquid hydrogen. I thought fast and told the workers, "Okay, dump that damned hydrogen. Bleed that tank dry." By now the hangar was filled with smoke and flames were visible above the model fuselage. The workers looked at me funny but did what I told them, and on that damp evening the cold hydrogen filled that hangar with a fog five feet thick. All we could see of one another were our heads. If it wasn't for the fire we might have had a good laugh. But the fire department noisily arrived at our hangar door and the next problem was that security didn't want to let them in. The firemen weren't cleared and this was a project above top secret. I couldn't believe the stupidity, but I took one of the security guys aside and said, "The whole place is under fog. They won't see what's on fire."

"What is inside?" the fire chief asked me.

"National security stuff. Can't tell you," I replied.

The firemen saw the fog and went running for their gas masks. Had they known we were playing around with liquid hydrogen so close to Burbank Airport, I'm sure they would have had my scalp, but they put out the fire in two minutes and went away, no questions asked. But Kelly was cranky with me. "Goddam it, Rich, why in hell did you use a damned wooden-framed stove? That was just asking for trouble." I told him he was nickel-and-diming me so severely on this project that a wooden-framed stove was all I could afford. He couldn't argue.

God knows how many hours I spent as part of a small team sitting in Kelly's office, reviewing all our data and trying desperately to pull a few range-extending tricks out of the bag. We knew damned well what the problems were. We missed our lift-over-drag ratio by 16 percent from what we originally had estimated, and our specific fuel consumption was disappointing too. We thought we would be able to achieve one-fifth the fuel consumption of a standard kerosene-fueled engine at Mach 2.5. Instead we were able to achieve only one-fourth the fuel consumption — not good enough to get us where we needed to go and back.

The only way to extend range was by improving fuel consumption, adding more fuel storage capacity and improving lift over drag to make the airplane fly more efficiently. We had done all that we could in each of these critical areas and were still a thousand miles short of our guarantee to the Air Force. Since I was his "expert" on the exotic fuel, Kelly asked for my opinion. I said, "Two thousand miles will only get us from Los Angeles to Omaha. We would have to land at a base that stored liquid hydrogen for us to refuel. Air-to-air refueling is out, so we would need strategically placed liquid hydrogen tank farms in Europe and Asia to refuel our airplane on its flights over Russia, leaving us with the nightmare problems of logistics and handling of a touchy, volatile fuel. Right now, we are having huge headaches shipping in our special fuel to our U-2 base in Turkey and that does not require special refrigeration and expert handling."

Kelly sighed and said he agreed with me.

He picked up the phone and called Secretary of the Air Force James Douglas Jr. "Mr. Secretary," he said, "I'm afraid I'm building you a dog. My recommendation is that we cancel Suntan and send you back your money as soon as possible. We don't have the range to justify this project."

It took several Pentagon meetings with Kelly before the Air Force reluctantly agreed with him. We had spent about $6 million in development costs and returned $90 million to the government. The punch line to the story is this: not long after the contract was canceled, the Soviets launched their *Sputnik 1* into orbit. The rocket engine that had carried it into space was hydrogen-fueled. The engine builder was Pyotr Kapitsa, who had been released from the gulag not to build an airplane but to launch *Sputnik*.

We had all guessed wrong.

But our exercise on the hydrogen airplane was not a total waste. General Dynamics was working on a hydrogen-powered rocket called Centaur, so we turned over to them all our cryostats and liquid hydrogen pumps. We in the Skunk Works had proved to ourselves that we could develop a large supersonic airplane and engine. Even before Powers was shot down, Kelly had determined that we would need to make a quantum leap in technology in order to keep our spy planes operational over Russia. Within months we would be planning a technological marvel called the Blackbird as successor to the U-2. Once again we'd be teamed with Dick Bissell and the CIA and challenged to produce a new miracle.

Postscript on the U-2

All through the escapades involving the hydrogen airplane, the main occupation inside the Skunk Works was maintaining the production line of new U-2s. Many Americans believe that the U-2 died the day that Powers was shot down. The CIA did in fact close down its secret bases overseas and come home, but we had sold more than twenty U-2s to the Air Force back in the late 1950s and more than twice that number since then, and there never has been a single day since that airplane became operational in 1956 that a U-2 isn't flying somewhere in the world on a surveillance operation for the blue-suiters, NASA, or the Drug Enforcement Agency. In fact, on more than one occasion over the years, the U-2 may have saved the world from thermonuclear war.

Although Eisenhower decreed the end to Soviet overflights after the Powers tragedy, the blue-suiters very soon began flying U-2s along the Soviet border using new technologies like side-looking radar, which could peer two hundred miles or more inside Russia, and carrying special electronic packages that could capture all the different military-band and radar frequencies used by the Soviet defense forces, helping us to build effective jamming devices in the event of hostilities. In many ways this electronic intelligence collecting was even more valuable than the photo-taking operations of old.

Around the time that Eisenhower left office in 1960, Dick Bissell and the CIA were teamed with Lockheed's Missiles and Space Company to put up the first spy-in-the-sky satellite. A satellite was locked into its orbit, but a U-2 could overfly any trouble spot on earth in a few hours. And we never stopped improving the airplane. We invented a novel interchangeable nose that could be unscrewed and unbolted in less than an hour and replaced for a particular mission — some noses carrying radar, some cameras, some air-sampling filters, some electronic devices recording radar and military traffic frequencies, some operating a special rotating camera that could follow the flight path of a Soviet-launched test missile. Other flights carried a heavy payload of electronic eavesdropping equipment. By monitoring their test missile firings we discovered the frequencies used by their missiles locking to a target. We used this information to counter with powerful jamming devices installed in our attack aircraft.

One unforeseen consequence of the Powers shoot-down was to make U-2 overseas bases a political hot potato for our host country allies allowing us to take off and land on their soil. The Russians were frothing at the mouth at the mention of the U-2, threatening dire reprisals — including air attacks — against any country hosting a U-2 base. As pressures built, both Japan and Turkey capitulated and asked us to fold our tents. To become more self-sufficient, we decided to develop an air-to-air refueling capacity for the U-2, extending its range to roam very far from home on spy missions. It would also allow us to do more low-level flying to avoid radar detection, which was very fuel inefficient. Refueling could extend the U-2's range to seven thousand nautical miles and fourteen straight hours of flying time, which pushed a pilot's fatigue beyond safety. The U-2s could receive nine hundred gallons of fuel from a KC-135 tanker in about five minutes. But most U-2 pilots agreed with our test pilot Bill Park, who flew the first extended-range mission and sighed, "Never again. My mind went numb ten minutes ahead of my ass."

The Army wanted to use the U-2 for battlefield surveillance. As for the Navy, Kelly had polished off a bottle of White Horse up at Edwards Air Force Base with a couple of old pal Air Force generals, one of whom had bet him that the Navy would never buy a U-2 because the airplane could not take off from a carrier deck. "You don't have the horses to get into the air," one general challenged Kelly, who got sore and told the general he didn't know what in hell he was talking about. Then the three of them staggered out of the officers club and paced out the length of a carrier deck on the main runway at Edwards. Later that day, they got a U-2 to take off. Kelly won his bet with ease. And so the Navy was interested in purchasing the U-2 for its own reconnaissance uses, including extended-range antisubmarine patrols.

Also through the CIA, we completed a deal with the Chinese nationalist government on Taiwan, selling them several U-2s for $6 million, along with the services of our ground crews and technicians. Kelly signed the contract for us, and Chiang Kai-shek signed for the Taiwanese, but the nationalist government had nothing to do with the operation except to provide pilots. The CIA was in charge and in control of the operation. They would be overflying Communist China from Formosa and were called Detachment H. This operation was one of the most tightly held secrets in the government. We began training six of their pilots. I remember briefing them on the U-2's propulsion system at Burbank in the summer of 1959. On one of the training flights from the Ranch, a Taiwanese pilot flamed out over Cortez, Colorado, and was forced to glide into a small county airport around dusk. The airport manager took one look and almost fainted. The airplane was one that he had never seen before — long and sleek with enormous wings.

The Stealth fighter being loaded with laser-guided bombs at its Saudi Arabian air base during Operation Desert Storm. *(Lockheed)*

The first production Stealth fighter at the Skunk Works assembly plant in 1980. *(Lockheed)*

Stealth fighter rolling out of its hangar at its secret base on the Nevada desert on the Tonapah Test Range. *(Denny Lombard and Eric Schulzinger)*

Above: A pair of Stealth fighters preparing for takeoff at their Tonapah base on the Nevada desert. *(Eric Schulzinger)*

Right: Model of the Stealth fighter undergoing radar testing at White Sands, New Mexico. The test results were so spectacular that the Air Force tamped on the tightest security lid since the atomic bomb. *(U.S. Air Force)*

Above: Colonel Al Whitley, Stealth wing commander. Photo was taken at Nellis Air Force base, Nevada, on April 1, 1995, after his return from Saudi Arabia. *(Eric Schulzinger)*

Below: Skunk Works crew celebrates the first successful test flight of the Stealth fighter on June 18, 198 Ben Rich is sixth from the right in the third row (to the left of man in dark T-shirt). *(Skunk Works)*

above: U-2 pilot Francis Gary Powers, who worked as a Skunk Works test pilot following his release from Soviet prison. *(Lockheed)*

right: A U-2R being assembled at Palmdale (Plant 42, Site 7) in the late 1960s. *(Lockheed)*

Above: Four downed Taiwanese U-2s on display in Peking public park in 1966. *(Life Magazine)*

Below: U-2 spy plane. *(Lockheed)*

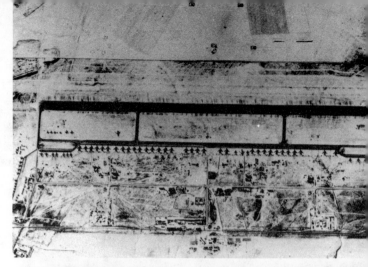

CIA pilot Marty Knutson's revealing 1957 photo of snowy Engles Airfield showing squadron of long-range Russian Bison bombers. *(Central Imaging Office)*

U-2 photo showing Soviet equipment at Cuba airfield during the Cuban missile crisis. *(Central Imaging Office)*

U-2 Nicaraguan overflight ordered by the Reagan administration, showing Soviet arms buildup in support of the rebel forces. *(Central Imaging Office)*

SR-71 Blackbird
taxiing to take off.
The world's fastest
airplane was also
a pioneer in
stealthiness. Note
angled tails that
greatly decreased
radar profile.
(Lockheed)

SR-71 Blackbird
streaking across
continental U.S.
(Lockheed)

Squadron of
Blackbirds
operating out of
Beale Air Force
base, near
Sacramento.
(Lockheed)

Blackbird
production line at
the Skunk Works.
In mid-1970s, they
produced a new
Blackbird every
month. *(Lockheed)*

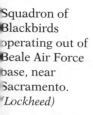

SR-71 Blackbird rolling down a highway en route to permanent display at the air museum at Robins Air Force base in Georgia. *(Jay Miller)*

D-21 drone being deployed from B-52 for round-trip spy mission over Chinese mainland. *(U.S. Air Force)*

B-52 carrying D-21 drone en route to launch off Chinese coast for spy mission over mainland. *(U.S. Air Force)*

D-21 spy drone tucked under wing of B-52 mothership. Drone would drop and ignite off Chinese coast for spy flight over mainland. *(Lockheed)*

Above: The Stealth Ship undergoing sea trials off the West Coast in the mid-1980s. *(Skunk Works)*

Above, right: The Stealth ship inside its secret hangarlike floating dock, where it was assembled and constructed. *(Skunk Works)*

Ben Rich and the F-117 Stealth fighter. *(Denny Lombard and Eric Schulzinger)*

The Stealth Ship, called the *Sea Shadow,* during sea trials. *(Skunk Works)*

Above: Lockheed's chief engineer Kelly Johnson talking to one of his pilots during a test flight in the 1940s. *(Lockheed)*

Right: Ben Rich and his old mentor, Kelly Johnson, pose in front of a new production model of the famous U-2 in 1983, following Kelly's retirement. *(Lockheed)*

Above: Ben Rich receiving the 1989 Collier Trophy, aviation's highest award, for the Stealth fighter. *(Lockheed)*

Below: The Skunk Works complex off the main runway of the Burbank Muncipal Airport in downtown Burbank, California. *(Lockheed)*

Then the canopy opened and out stepped an alien in a space suit, with only almond-shaped eyes visible through his visor, who ran to him, shouting in very garbled English, "Quick. Get gun. Guard plane. Very, very secret."

The Taiwanese squadron, which became known as the Black Cats, was a joint CIA-Taiwanese operation, flying from Taoyuan airfield, just south of Taipei. These U-2 flights over Red China lasted for more than fourteen years, from late in 1959 until 1974, when President Nixon finally put a stop to them in deference to his new diplomatic opening to the People's Republic. But especially during the early 1960s, the overflights were considered by the intelligence community to be extremely urgent. We needed hard information on Chinese nuclear and missile development. The Pentagon was particularly eager to learn how the Sino-Soviet split was affecting China's military capacity and its weapons procurements.

The flights were much more grueling and dangerous than the Soviet overflights — typical eight- to ten-hour missions calling for a three-thousand-mile flight over hostile territory practically from takeoff to landing. To reach the highest-priority targets of nuclear test sites in northwestern China and the Chiuchuan intermediate ballistic missile range in Kansu province meant flying twelve-hour round-trips. Over the years, as Chinese ground-to-air missile defenses improved, the Taiwanese took a pounding. Four U-2s were shot down and their pilots lost. During the sixties, the remains of those downed airplanes were put on display in downtown Peking, and the overflights so enraged the Communist Chinese, they offered $250,000 in gold to the Taiwanese pilot who would defect with a U-2 to the mainland. And no wonder. The intelligence acquired by these flights was so revealing that U.S. experts were able to accurately predict when the Chinese would finally test their first nuclear weapon in October 1964.

Back in Burbank, we did what we could to help cut down the U-2 losses. We developed improved electronic counter-

measures (ECM), calculated to confuse Chinese radar operators working their SA-2 ground-to-air missile systems. On radar screens the U-2 would present a false display so that the
missile would be launched in the wrong piece of sky. Our
ECM package was bulky and heavy and cost around two hundred gallons of fuel-carrying capacity, cutting into range and
altitude performance.

Some of the more distant nuclear test sites near the Tibetan
border were out of range of the Taiwan-based U-2s. To cover
these targets the agency flew from dirt landing strips in India
and Pakistan on an ad hoc basis. In fact, three months before
Powers was shot down over Russia, a CIA pilot flew from a
secret base in Thailand against Chinese nuclear facilities. The
U-2 dropped a javelin spike that we had dreamed up that contained special miniature seismic sensors to record an expected thermonuclear bomb test. Unfortunately, we never got
any data back and never learned why. But the pilot on that
mission was forced to crash-land short of his base in Thailand
and came down in a rice paddy. He was able to negotiate a
deal with the village headman: the villagers helped him to cut
up the U-2 and put the pieces aboard oxcarts and haul it to a
clearing, where a CIA C-124 landed the next day and took him
and his plane out. In return, the agency paid the headman five
hundred bucks to build a schoolhouse. Gary Powers should
have been so lucky.

Other Voices
Buddy Brown

I was just a dumb twenty-three-year-old fighter jock, which is
exactly what the Air Force was looking for back in 1957. All
they told me was "How would you like to fly at very high altitude in a pressure suit?" I immediately thought, Rocket ships!
Buck Rogers! Count me in. I was shipped down to Laughlin Air
Force Base in Del Rio, Texas, on the Mexican border, way out

of sight, which is how the Air Force wanted it, because it wasn't until the 1962 Cuban missile crisis that the world learned the Air Force was flying U-2s.

We had twenty airplanes there and Air Force instructors to check us out, but we had a lot of fatalities. The U-2 was strictly a one-seater. The first time you flew it, you soloed, ready or not. We did a lot of landing pattern and takeoff practicing, and got up to sixty thousand feet to get the feel of our pressure suits. It was a very tricky airplane and we had a lot of fatal pilot errors. One guy was killed flying over his house, while showing off for his wife and two little boys. He banked too low and slammed into a hill. Another time the squadron commander was forced to eject when his flap switch stuck and he lost his tail and we didn't have an ejection seat. So he jumped out, making the highest bailout ever — a record fifty-five thousand feet — and was very badly hurt. Another time I watched a guy nose in on landing and kill himself. I shit because I had to fly next.

My first assignment was the most dangerous flying I had ever done — by far. I flew out of Alaska in what was officially called the High Altitude Sampling Program. That meant flying into the drifting radioactive clouds following Soviet and Chinese nuclear tests. Up there on polar flights when the sky was crystal clear and you could see the curvature of the earth, you'd be able to spot the nasty-looking iodine cloud drifting from god knows how many miles off. And we'd fly right into it. That program was entirely Air Force, and every bit as important as the agency flights over Russia. We flew for the Defense Atomic Support Agency, which collected our six bottles of gaseous samples of particulates after each flight and rushed them back to Washington for laboratory analysis.

They could tell by debris samples carried in the wind whether the Chinese exploded an air or ground burst, what part of the country it was set off in, how advanced their trigger and weapon were just by the materials that vaporized. And we always knew their tests from our own because we placed a tiny metal object in our nuclear devices that left an unmistakable signature on a spectroscope. We figured we were pretty safe from radiation hazards while insulated in a pressure suit, but we were naive about the dangers in those days. The most penetrating radiation was believed to decay so quickly that by the time we flew into a cloud of gases and suspended debris, the risks were supposedly minimal. We wore radiation badges. Still, every so often an aircraft landed very damned hot and had to be isolated and washed down and the pilot spent the night in hospital as a precaution. As far as I know, no one was the worst for it.

We flew these sampling missions every Tuesday and Thursday, in conjunction with other blue-suiters flying U-2s in Puerto Rico and Argentina, taking an opposite route from us. So we had one north and one south mission, and in that way we were able to sample half the globe per mission.

I flew some sampling missions out of Laverton, which was the Australian version of Edwards Air Force Base, flying toward Antarctica. I was more fearful then than I was later flying U-2 reconnaissance flights in combat over Vietnam. The reason was the extreme weather. You'd last two seconds if forced to bail out in those awful temperatures. And you'd last five minutes on the ground. The distances were so vast, there was no way to be rescued in time.

I flew at a time when the Chinese were exploding a lot of nukes, so I got used to ten-hour missions. I drank a pint and a half of orange juice through my feeding hole in my helmet, but even so, after a long flight my fingernails were so brittle from body dehydration that I could just crack them off. We also worried that ozone from so much high-altitude flying would rot our teeth. Maybe that was an old wives' tale, but we all worried about it; I got the base dentist to make me a set of rotten-looking greenish false teeth to wear over my real teeth at base parties.

I also flew a lot of what we called peripheral missions, flying just outside the borders of the Soviet Union or China, collecting intelligence. All I had to do was throw a switch and recorders on board would collect the bad guy's radar frequencies and signals, and monitor everything. I remember one particular mission, code-named Congo Maiden, where we had five U-2s up there at the same time in the northern part of the Soviet Union. We carried on board an ECM package called a System 12, so you knew when you were being picked up by Soviet radar by hearing pings in your headset. Tightened my sphincter for sure.

I flew Vietnam missions out of Okinawa as early as 1960. I flew over the Plain of Jars and watched the French get their butts kicked by Uncle Ho. Then, in '62, the Russians took a few shots at me with SA-2s during the Cuban missile crisis. Didn't come close thanks to my black box in the tail that jammed effectively. So I'm a believer.

But that was inconsequential compared to another blue-suiter U-2 pilot, Major Chuck Maultsby, who was flying out of Alaska on a routine sampling mission right at the height of the 1962 Cuban missile crisis. His mission took him over the North Pole in the middle of the night, and when he turned to return to Alaska, he took the wrong south heading and wound up flying deep into Soviet territory. The Russians picked him up right away and thought SAC was coming in the back way to nuke them and start World War III. We monitored them scrambling jets against Chuck. He could see the contrails of dozens of fighters trying to reach his altitude and shoot him down. Finally, President Kennedy got on the hot line with Khrushchev and told him we have a lost U-2 pilot over your country on a weather mission, and he is not — repeat, not — a hostile aircraft. Maultsby had no direct radio communications, only a passive HF receiver that allowed him to listen. Someone on the tanker that had refueled him got on the horn and informed Chuck that it was sunrise over Alaska and suggested he turn his airplane 90 degrees until he saw light, then fly in that direction. Chuck obeyed and headed for the western tip of Alaska, where he was met by a couple of our F-106s that escorted him to base. He had made the longest U-2 flight ever — about fifteen straight hours and ran his fuel down to zero, flamed out, had to deadstick in with his face mask all frosted over.

The CIA had been covering Cuba with U-2 flights for years. And then, in August 1962, they hit pay dirt and came up with the pictures that showed the Russians were planting ballistic missiles right next door, SS-4s and SS-5s. When Kennedy was shown the site constructions, he asked, "How do we know these sites are being manned?" They showed Kennedy a picture taken from 72,000 feet, showing a worker taking a dump in an outdoor latrine. The picture was so clear you could see that guy reading a newspaper.

The first thing Kennedy did was to step up the flights. The second thing he did was to take the agency off the case and put in us blue-suiters in their place. If a guy was shot down, he wanted it to be a military driver, not a CIA employee. So I was one of eight Air Force guys who took over the Cuban overflights during the crisis. We flew out of McCoy Air Force Base in Florida, three or four missions a day. Since our missions were relatively short, we carried less fuel and so we could climb higher than usual, which was good because some of these missions got hairy.

On October 27, one of our guys, Major Rudy Anderson, got nailed when an SA-2 missile, fired from a Cuban naval base at the eastern end of the island, exploded above and to the rear. Shrapnel blasted into Rudy's canopy and blew holes into him. It was standard procedure to brief a primary and backup mission pilot for each day's mission. The morning Rudy was hit by a SAM, I was flying the primary mission area, while Rudy was scheduled to fly the backup mission if my area was weathered in. As it turned out, my area was completely socked in with clouds, so Rudy flew the backup mission and got hit. One of the more awful aspects to this tragedy happened during a training accident earlier in the year. A pilot named Campbell was killed during a refueling exercise back in California. A garbled message got back to Edwards base control tower that Anderson was the pilot killed and everyone rushed over to Rudy's house to comfort his wife, Jane. Well, you can imagine the impact on Jane until the phone rang and she heard Rudy's voice and then damned near fainted away. Then she was forced to go through the same shit the second time only eight months later — but this time for real.

After Rudy was shot down, we got the word that Kennedy had warned Castro and Khrushchev that if another reconnaissance airplane was shot down, we would stage an all-out bombing attack against these installations. The rumor was he was prepared to nuke the island. If we heard that rumor, figure the Cubans did too.

I was selected to fly to Homestead Air Force Base, in Florida, and brief President Kennedy on the Cuban missions. When I was introduced to the president, he smiled and remarked, "Major Brown, you take damned good pictures."

In late 1963, we began launching U-2s from U.S. aircraft carriers, having developed a workable tailhook. In May 1964, the U-2 took off from the USS *Ranger* to monitor French nuclear tests in an atoll in French Polynesia, but only after one of our test pilots, Bob Schumacher, crashed while landing on deck. We had the airplane fixed and flying by the next morning. The target of the operation was Mururoa Atoll, a part of French Polynesia. We monitored all of their testing, and the French never knew we were observing them. The flights were secret, and the carrier crew had to go below deck when the bird took off and landed. The agency painted on its tail "Office of Naval Research," just in case it was forced to crash-land in French territory. The photographic evidence acquired by the overflights revealed that DeGaulle's government would be ready for full-scale nuclear weapons production in a year.

During the Vietnam War, we launched gliders from our U-2s as decoys — a Kelly Johnson idea. The gliders carried tiny transmitters that fooled the North Vietnamese missile batteries into thinking they were actually B-52 bombers or fighter-bombers. So for $500 a decoy we forced them to launch missiles costing thousands of dollars.

Other Voices

James R. Schlesinger
(Director of the CIA 1973; Secretary of Defense, 1973–75)

As secretary of defense, I confronted my own version of a Cuban missile crisis scenario in the mid-1970s, when I suddenly found myself under enormous political pressure and the U-2 came to my rescue and bailed me out. This happened during the Ford administration, in the spring of 1975, a period during which the Soviets were aggressively establishing bases and influence in northeastern Africa, in places like Somalia, Angola, and Uganda. Henry Kissinger, then secretary of state, was pushing aggressively for detente with the Soviets. He and I were on opposite ends of a tug of war about establishing an American naval base in the Indian Ocean on the British-owned island of Diego Garcia. Kissinger was adamantly opposed to building such a base and had a lot of powerful support for his position in Congress. Democratic Majority Leader Mike Mansfield urged that the entire Indian Ocean region remain "a zone of peace" that would preclude us from operating there. The dispute with Congress over that base was endless. The Russians also screamed loudly about the provocation of an American naval installation in the Indian Ocean, even though they were crawling all over the place, aggressively extending their influence throughout the region. We had good intelligence on what they were up to in Somalia and Uganda, which were pretty much under their domination.

In April, spy satellite photos landed on my desk showing that the Soviets had constructed a missile handling and storage facility at the Somalian port of Berbera, commanding strategic approaches to the Red Sea, which would be a depot for storing Styx missiles used by the Soviet fleet in the Indian Ocean. These were missiles fired against other ships. The pictures provided proof of a Soviet military buildup in the area, but I was stymied by a blanket injunction against any public disclosure of satellite photography, extending even to members of Congress. In those days we didn't admit that spy satellites existed, so I could not release the pictures, especially to make a political point. Instead, I ordered the Air Force to schedule a U-2 flight over the Berbera installation and provide overhead photos that I could make available to the press. The photos taken by the U-2 were superb, and I decided to go public and announced that the Soviets had begun storing missiles in Somalia. I knew that my announcement would fire a lot of angry skepticism in my direction, among detente proponents on the Hill as well as among some in the press, who heaped scorn on the Pentagon, claiming we were eager to sabotage detente and using scare tactics to overcome congressional opposition to a U.S. base in the Indian Ocean. The Russians and the Somalis vigorously denied my accusation. The Russians claimed they were only building a meat-packing plant at Berbera, and nothing more. Kissinger was concerned that I was about to upset his detente policy, so he was not enamored about having me release the U-2 pictures to the press to prove my contention. To be frank, he was rather infuriated with me over the entire episode, especially when I showed the U-2 pictures to the Senate Armed Services Committee and gave copies to the *New York Times*, which ran a picture in early June. The Russians called the picture "a mirage," intended to win support for a larger Pentagon budget. But for me the release of those U-2 photos became a jolly good episode. Once again, overhead photography caught the Russians trying to upset strategic balances just as they had in the Cuban missile crisis by secretly extending their military capabilities on friendly shores. But before that summer ended, the U-2 pictures had nailed our case: the Somali government backed off its futile denials and, trying to save face and win congressional support for drought aid, actually invited us into Berbera to build a naval supply installation of our own.

In August 1970, Henry Kissinger arranged for two U-2s to monitor the unsettled Middle East buffer zone between Israel and Egypt. And in April 1974, after twenty years, the CIA ended its aviation activities and turned over all its twenty remaining U-2 aircraft to the Air Force. In more recent years the airplane has seen service monitoring the oil leak in the Santa Barbara channel, the Mount St. Helens eruption, floods, topography, earthquake and hurricane damage assessments, and by drug enforcement agencies to monitor poppy fields around the globe. The DEA (Drug Enforcement Administration) was involved in a test project on the U.S.–Mexican border in the late 1970s to test infrared film filter combinations on poppy fields photographed from high altitudes. Every growing thing has its own infrared signature, and the agents wanted to discover how poppies photographed at various stages in their growth cycle; photo interpreters could tell how close to harvesting a particular field was. The field in question was in Yuma, Arizona, specially cultivated under the DEA's supervision, using fugitive Mexican poppy planters. A U-2 would overfly the field at various stages in the growth cycle and photograph it. Finally, the agency, after conferring with the Mexican planters, ordered a last flight for photos showing poppies ready to harvest. The U-2 flew over the field, as scheduled, only to discover the poppy field had been swept clean: the workers had harvested the crop the night before and slipped back into Mexico. The first U.S. government–subsidized and grown heroin was probably on the streets a few weeks later.

When we in the Skunk Works first built the U-2, we thought it would be in production for about eighteen months, but it is still in service. During Operation Desert Storm, the U-2 over-flights monitored Iraqi tank movements, and its side-band radar proved effective in revealing the presence and configuration of enemy mine fields. In January 1993, when the outgoing Bush administration decided to bomb Saddam Hussein's missile batteries in the southern "no-fly" zone, the U-2 was once again providing the vital intelligence data preliminary to the bombing. On the day before the bombing raid, I received a call at home from an official of the CIA. "Ben," he said, "we just got a call from President-elect Clinton. He wants to know the altitude of the U-2. No one at this end is sure, so I thought I'd go straight to the horse's mouth."

"Tell the president-elect that our bird flies at seventy thousand feet." And I said it with pride.

9

FASTER THAN A SPEEDING BULLET

The Blackbird, which dominated our work in the sixties, was the greatest high-performance airplane of the twentieth century. Everything about this airplane's creation was gigantic: the technical problems that had to be overcome, the political complexities surrounding its funding, even the ability of the Air Force's most skilled pilots to master this incredible wild horse of the stratosphere. Kelly Johnson rightly regarded the Blackbird as the crowning triumph of his years at the Skunk Works' helm. All of us who shared in its creation wear a badge of special pride. Nothing designed and built by any other aerospace operation in the world, before or since the Blackbird, can begin to rival its speed, height, effectiveness, and impact. Had we built Blackbird in the year 2010, the world would still have been awed by such an achievement. But the first model, designed and built for the CIA as the successor to the U-2, was being test-flown as early as 1962. Even today, that feat seems nothing less than miraculous.

KELLY JOHNSON'S disappointment over our failure to produce a workable hydrogen-powered airplane had him pouting for a day or two, but he quickly recovered and began lobbying the CIA for a new spy plane to fly over Russia that would be a quantum leap over the U-2 in every way. He assembled the small group of us who had worked on the hydrogen plane and had us brainstorming ideas for new designs and approaches for an airplane that used conventional engines and fuel but still could outrace any Russian missile. "It makes no sense," he said, "to just take this one or two steps ahead, because we'd be buying only a couple of years before the Russians would be able to nail us again. No, I want us to come up with an airplane that can rule the skies for a decade or more."

At that point, in April 1958, the U-2 overflights of Russia were in their second year and going well. In fact, it would be two more years before Francis Gary Powers was shot down, and the high priority at the Skunk Works was Operation Rainbow, our attempt to lower the U-2's radar cross section. But Kelly declared the U-2 doomed. The Russians had made it a matter of national honor to find a way to stop U-2 overflights and were investing billions of rubles in rushing to develop a missile system to do so. Dick Bissell shared Kelly's glum outlook for the U-2's future and encouraged him to begin sketching out a successor spy plane. "We'll fly at ninety thousand feet, and jack up the speed to Mach 3. It will have a range of four thousand miles," Kelly told a group of us. "The higher and faster we fly the harder it will be to spot us, much less stop us."

I didn't know about Richard Bissell of the CIA, but Ben Rich of the Skunk Works reacted to Kelly's idea with jaw-dropping disbelief. He was proposing to build an airplane that would fly not only four times faster than the U-2 but five miles higher — and the U-2 was then the current high-altitude champion of the skies. A Mach 3 airplane was 60 percent faster than the maximum dash capability of our top-performance jet fighter. Experimental rocket airplanes had flown at blinding Mach 3 speed using powerful thrusters for two or three minutes at a time until fuel ran out. But Kelly was proposing an airplane to *cruise* at more than three times the speed of sound, that could fly coast to coast in less than an hour on one tank of gas.

Kelly's audacious idea would probably not have been taken

seriously by the CIA coming from anyone other than the boss
of the Skunk Works. After all, in 1954 we had built the F-104
Starfighter, the world's first Mach 2 fighter. So a Mach 3 airplane seemed a logical extension of our skills. However, there
was a Grand Canyon–size gulf between designing an airplane
like the F-104 that could kick in its afterburners on takeoffs
and in dash modes lasting a minute or two, and designing an
airplane whose "normal" cruising speed was nearly twice as
fast as the fastest fighter's dash speed. On afterburners, a
fighter was burning fuel at a rate four times faster than at
cruise speed, so afterburners were saved for combat threat
situations — escaping flak after a bombing run or outflying
missiles or dogfighting MiGs. We were proposing to fly whole
missions on afterburners. The technology confronting us was
so far beyond anything on the drawing boards of any other
aerospace company in the world that we might as well have
been proposing commuter rocket service between the moon
and the outer ring of planets.

For openers, to be able to fly sustained at such heights and
speeds would require radical departures in how we designed
and built propulsion systems.

"Rich," Kelly said, turning my way, "I'm making you program manager for the propulsion system." He ignored my
stunned expression. "How hot do you suppose the airplane
will get at Mach 3 in sustained flight?" he asked me. "Somewhere between a blowtorch and a soldering iron, I guess," I
replied when my voice returned. He nodded. "Probably
around nine hundred degrees at the nose," he said. "Just imagine that kind of thrust! You're the lucky one. You'll at least
have known laws of physics to guide you. The rest of us are
going to have to do some fancy stretching to find out what can
work. We start from scratch as if we are building the first airplane, just like the Wright brothers."

If I had been older and smarter, I would've run for the nearest exit. I had to produce a propulsion system more efficient

than any other ever designed. I was then only a thirty-two-year-old fledgling, still on probation to prove my worth as a propulsion and thermodynamics engineer among many of my senior colleagues. But I was cocky enough to shrug off Kelly's challenge and think, A Mach 3 airplane! Why in hell not? Kelly surrounded himself only with the kind of can-do guys that made American aerospace technology preeminent. To him, the word "impossible" was a gross insult.

Kelly promised to deliver the world's first Mach 3 airplane to the CIA only twenty months after we signed a contract. That also seemed to me, in my pathetic innocence, a reasonable deadline. After all, it had taken us only eight quick months to deliver the first U-2. Had I really thought about it, in complexity the U-2 was to the Blackbird as a covered wagon was to an Indy 500 race car.

To action-oriented guys like Bissell and Kelly, President Eisenhower often moved too cautiously. In pique, they referred to him as "Speedy Gonzales," while being forced to cool their heels for weeks or months awaiting Oval Office decisions, whether for approving a particular U-2 mission over Russia or signing off on a new spy plane project. Kelly's airplane was bound to cost millions, and would be a tough sell. The president was already spending a billion dollars in covert funds on the Agena rocket that would boost our first spy satellite into orbit. Bissell was in charge of that program, too, and the first twelve test firings had all been failures. Lockheed's Missiles and Space Company in Sunnyvale, California, had that contract, and Bissell asked Kelly to evaluate and reorganize their operation. Kelly set up a mini Skunk Works and, coincidentally or not, the thirteenth test shot was a success. But spy satellites had distinct limitations: their pictures in those days were not very sharp, and their orbits were fixed, so the Russians would learn to hide secrets before each scheduled overflight. By contrast, a spy plane operated on no fixed schedules, could loiter in areas of interest, and could overfly tension spots within hours. Our photography was vastly superior to a satellite's.

Ike tremendously valued the U-2 photo takes but continually worried about the consequences of a shoot-down. He was attracted to the satellite alternative because he felt it was a less aggressive and threatening way to obtain overhead intelligence. Nations would learn to live in the age of satellites, but a spy plane flight would always be regarded as a provocative and aggressive violation of a country's territory. So Bissell wisely decided to seek the backing of Ike's two most influential technology advisers — Dr. James Killian, the president of the Massachusetts Institute of Technology, and Dr. Edwin Land of Polaroid, who chaired the presidential advisory panel on aerial espionage and was the godfather of the U-2 program. In May 1958, Kelly flew to Cambridge to meet with Dr. Land and his associates. At that first meeting he was amazed to learn that the Navy had its own Rube Goldberg blueprint for a high-flying spy plane. Theirs would be a ramjet, lifted high into the stratosphere by a balloon. At 100,000 feet, the pilot would release the balloon, light booster rockets to get his ramjet started, then roar up to 155,000 feet. A Navy commander presented this unique idea to the panel while Kelly sat scribbling figures on a pad. "By my calculations," Kelly told the group, "in order to lift that ramjet, the Navy's balloon would have to be over one mile in diameter. Gentlemen, that's one hell of a lot of hot air."

A more serious proposal came from Convair, which had also been solicited by Bissell for ideas on a high-flying, high-speed spy plane. They had built the B-58 Hustler bomber, a highly regarded Mach 2 tactical strike airplane, which they presented as a "mothership" that would launch a piloted rocket plane that supposedly could reach 125,000 feet at Mach 4. The piggyback launch concept interested Land, but as months passed and the idea was further refined and tested, it became increasingly obvious that the B-58 could not go supersonic while carrying a smaller bird under its belly. Kelly was also skeptical about whether Convair's plan could produce a reliable photo platform, and he wasn't shy in passing along his doubts to his CIA friends.

Over the next year Kelly shuttled back and forth to Washington, meeting with Bissell, Land, and other panel members, offering them our latest designs and radar test data, often returning dejected by rumors that Convair's proposals promised better performance and radar-cross-section data than ours, even though our first preliminary design drawing looked terrific. It was designated A-1, and showed a striking single-seat, two-engine airplane — a long, sleek, bullet-shaped fuselage with rounded inlets on big engines mounted on the tip of small delta wings that were two-thirds of the way back on the fuselage. One look and even a schoolboy would realize that this bird was designed for blazing speed. But the president was less interested in performance and more intent on pushing for the lowest radar cross section possible. It wasn't that he just didn't want to get us shot down — he didn't want the Russians to know we were even up there.

Kelly argued with Washington that our tremendous height and speed advantages were the most potent factors in making us difficult to detect, but the White House and the CIA were not mollified. So we decided to apply radar-absorbing ferrites and plastics to all the airplane's leading edges — a first in military aviation. We kept the twin tails as small as possible and decided to try to construct them entirely with radar-absorbing composites — a significant technological breakthrough if we could actually do it. But "hiding" this airplane seemed impossible. The tremendous heat generated in supersonic flight made infrared detection inevitable. How do you hide a meteor? Our Mach 3 airplane would streak across the sky like a flaming arrow.

About six months into the design phase I could see discouragement clouding Kelly's big round face. Our design was now numbered A-10 and we still were not achieving lower radar-cross-section results than Convair, according to Dick Bissell. So, in late March 1959, we began a series of almost around-the-clock brainstorming sessions to review all our previous work and to somehow find a design that would elude Soviet radar. But it seemed fruitless, and Kelly invited Bissell and a couple of agency radar experts to Burbank for what was to be a showdown "where we stand" meeting. He asked me and two others from the design team to sit in and lend him moral support.

The meeting was tense and somber; Kelly was typically candid. "We've put in six months of intensive design and study, and by God, we know what we're doing, but we will never get to the point where the president will be happy with the results. I'm convinced that current improvements in Russian radar will allow them to detect any airplane built in the next three to five years. Radar technology is far ahead of antiradar technology, and we're just going to have to live with that fact. We'll never achieve the zero degree of visibility the president seems so stuck on. That technology is way beyond what we know how to do at this point. Maybe Convair can deliver it for you. But we can't."

Not much more was said. And when the CIA officials left, Kelly said to us, "Well, boys, I think we're out. Ike wants an airplane from Mandrake the Magician."

But later he took me aside. "Keep after this, Ben. Maybe Land or someone else will get Ike to see the light."

We kept working mostly because it was an unusually slack period at the Skunk Works, without too many other competing distractions. By design A-11, in May 1959, we felt we had scored a breakthrough in dramatically lowering the radar cross section of the aircraft. One of the structural designers presented the idea of modifying the bullet-shaped fuselage by adding a chine, a lateral downward sloped surface that gave the fuselage an almost cobralike appearance. Now the underbelly of the airplane was flat, and the radar cross section had magically decreased by an incredible 90 percent.

By July, we decided to lay out a final revised drawing of the entire airplane making full use of the new chine configuration. In those days I shared an office with four others working on the new airplane — aerodynamicist Dick Fuller, two others who did performance and stability control, and my own sidekick in propulsion, a brilliant twenty-four-year-old Caltech grad named David Campbell, an aerothermodynamicist. (Dave was destined for true greatness, but only two years later, during his daily two-mile jog, he dropped dead from a massive coronary; he was only twenty-six years old.)

I was separated by a connecting doorway from the office of four structures guys, who configured the strength, loads, and weight of the airplane from preliminary design sketches. They put skin and muscle onto the original design concept.

After lunch one blazing summer afternoon, the aerodynamics group in my office began talking through the open door to the structures bunch about calculations on the center of pressures on the fuselage, when suddenly I got the idea of unhinging the door between us, laying the door between a couple of desks, tacking onto it a long sheet of paper, and having all of us join in designing the optimum final design to make full use of the chines. My object was simple. I said, "We're never going to get this design a hundred percent right. We could play around forever. But I think we now know enough to nail it down at eighty percent. And that's plenty good enough."

One of the participants later wisecracked that it was like the Russian and American soldiers joining up on the banks of the Elbe River during the last days of World War II. It took us a day and a half; Ed Baldwin did the basic design and Ed Martin the systems. Henry Combs and Ray McHenry did the structures. Merv Heal figured the weights and Lorne Cass the loads. Dan Zuck designed the cockpit, and Dave Robertson handled the fuel system requirements. Dave Campbell and I weighed in with propulsion and Dick Fuller and Dick Cantrell with the aerodynamics. Everyone chipped in with changes and modifications from previous designs. The airplane weighed 96,000 pounds without fuel — keeping it light to maximize fuel consumption and minimize cost — and was 108 feet long with an extremely thin double delta wing attached at mid-fuselage. The wing edge was designed so razor-thin that it could actually cut a mechanic's hand. We took the long sheet of paper to Kelly Johnson and unrolled it on his desk. We told him, "Kelly, everything is now exactly where it should be — the engines, the inlets, the twin tails. This is probably as close to the best we can come up with."

It was our twelfth design, number A-12, which would later become its official CIA project designation. Kelly took the design and ran with it to Washington. Throughout midsummer 1959, he shuttled back and forth to CIA headquarters in Langley, Virginia, nearly a dozen times, meeting with Bissell, Land, Allen Dulles, and others, noting at one point in his private journal, "There is a good deal of concern that Speedy Gonzales [Ike] will cancel. Too expensive."

But the deal was finally nailed on August 28: "Saw Mr. Bissell alone. He told me that we had the project and that Convair was out of the picture. The agency accepts our conditions that our method of doing business will be identical to that of the U-2. Mr. Bissell agreed very firmly to this latter condition and said that unless it was done this way he wanted nothing to do with the project either. He and Allen Dulles stated following conditions: (1) We must exercise the greatest possible ingenuity, an honest effort in the field of radar. (2) The degree of security on this project is, if possible, even tighter than on the U-2, and (3) We should make no *large* commitments, *large* meaning in terms of millions of dollars."

We were being funded to build five A-12 spy planes over the next two years at a quoted price of $96.6 million.

God help us, we were in business.

The CIA code-named the project Oxcart, an oxymoron to end all: at Mach 3, our spy plane would zip across the skies faster than a high-velocity rifle bullet.

Kelly had sold the idea brilliantly, but now it was up to us peons to deliver the goods. One of the great strengths of the place was the combined experience of Kelly's most senior and trusted engineers and designers, who, among other attributes, were walking parts catalogs. But suddenly we were all operating in the dark, struggling by trial and error, like Cro-Magnons trying to look beyond the cooking fire to the first steam engine.

All the fundamentals of building a conventional airplane were suddenly obsolete. Even the standard aluminum airframe was now useless. Aluminum lost its strength at 300 degrees F, which for our Mach 3 airplane was barely breaking a sweat. At the nose the heat would be 800 degrees — hotter than a soldering iron — 1,200 degrees on the engine cowlings, and 620 degrees on the cockpit windshield, which was hot enough to melt lead. About the only material capable of sustaining that kind of ferocious heat was stainless steel.

For security and convenience, Kelly kept those of us working on his airplane jammed together in one corner of our old Building 82, a remnant of the bomber factory from World War II, in which we had built the U-2 and the F-104 Starfighter before it. From the original four he had approached on this project, we had now grown to a modest fifty or so, seated at back-to-back desks, where, like the early U-2 days, privacy surrendered to incessant kibitzing, teasing, brainstorming, and harassment. Some wag hung the sign PRIVACY SUCKS. My three-man thermodynamics and propulsion group now shared space with the performance and stability control people. Through a connecting door was the eight-man structures group, who designed the strength and load characteristics of the airplane. Their "dean" was that irascible genius Henry

Combs, who had been with Kelly since World War II and helped him build the classic two-engine P-38 Lightning interceptor. Henry and I could have reached through the doorway and shaken hands. And, of course, he relished offering his unsolicited advice and opinions to the young whippersnapper running thermodynamics and propulsion.

"Ben Rich," he teased, "how in hell do you propose getting our stainless-steel monstrosity up to speed at Mach 3? You'll need inlets the size of the Holland Tunnel." Then he chuckled sardonically.

Of course, Henry was no happier with the prospect of building a steel airplane than I was. More weight meant more internal support structure, more fuel, less range, and less altitude. Back in 1951 he had recommended to Kelly that we use a rare alloy called titanium for the white-hot exhaust nozzles on the afterburner of the supersonic F-104 Starfighter. So Henry Combs now was mulling the pluses and minuses of building the world's first titanium airplane. It would be a huge risk. On the positive side, titanium was as strong as stainless steel but only half its weight and could withstand blast-furnace heat and tremendous pressures. Titanium's tensile strength would allow us to make our wings and fuselage paper-thin. But to build a high-performance aircraft out of such an unproven exotic material was inviting potential disaster. "Unpredictability is a guarantee that we'll be in the soup on this one from start to finish," Henry predicted dourly. I knew he was right. Meanwhile, Kelly was already pondering the titanium idea himself. "Any material that can cut our gross weight nearly in half is damned tempting," Kelly told Combs, "even if it will drive us nuts in the bargain."

Only one small U.S. company milled titanium, but sold it in sheets of wildly varying quality. We had no idea how to extrude it, push it through into various shapes, or weld or rivet or drill it. Drilling bits used for aluminum simply broke into pieces trying to pierce titanium's unyielding hide. This exotic alloy would undoubtedly break our tools as well as our spirits. At one of our daily seven a.m. planning sessions in Kelly's office, I volunteered some unsolicited advice about how we could use a softer titanium that began to lose its strength at 550 degrees. My idea was to paint the airplane black. From my college days I remembered that a good heat absorber was also a good heat emitter and would actually radiate away more heat than it would absorb through friction. I calculated that black paint would lower the wing temperatures 35 degrees by radiation. But Kelly snorted impatiently and shook his head. "Goddam it, Rich, you're asking me to add weight — at least a hundred pounds of black paint — when I'm desperately struggling to lose even an extra ounce. The weight of your black paint will cost me about eighty pounds of fuel." I said, "But, Kelly, think of how much easier it will be to build the airplane using a softer titanium, which we can do if we lower the heat friction temperatures on the surface. Adding a hundred pounds is nothing compared to that."

"Well, I'm not betting this airplane on any damned textbook theories you've dredged up. Unless I got bad wax buildup, I'm only hearing you suggest a way to add weight."

Overnight, however, he apparently had second thoughts, or did some textbook reading on his own, and at the next meeting he turned to me as the first order of business. "On the black paint," he said, "you were right about the advantages and I was wrong." He handed me a quarter. It was a rare win. So Kelly approved my idea of painting the airplane black, and by the time our first prototype rolled out the airplane became known as the Blackbird.

Our supplier, Titanium Metals Corporation, had only limited reserves of the precious alloy, so the CIA conducted a worldwide search and, using third parties and dummy companies, managed to unobtrusively purchase the base metal from one of the world's leading exporters — the Soviet Union. The Russians never had an inkling of how they were actually contributing to the creation of the airplane being rushed into construction to spy on their homeland.

Even before the first titanium shipment arrived, many of us were already worrying that building this particular airplane might just prove too difficult, even for the Skunk Works. Wind tunnel tests of our mock-up amazed us all by indicating that, at Mach 3, intense friction heating on the fuselage would actually stretch the entire airframe a couple of inches! The structures people struggled like medieval alchemists to find rare and exotic metals that could withstand such blowtorch temperatures. They recommended that the hydraulic lines be of stainless steel; for the ejector flaps they found a special alloy called Hastelloy X; and they recommended making our control cables out of Elgiloy, the material used in watch springs. Plumbing lines would be gold-plated since gold retains its conductivity at high temperatures better than silver or copper. Kelly just fumed watching our materials costs rocket into the stratosphere.

There was simply no way to cut any corners. We discovered that there was no off-the-shelf, readily available electronics — none of the standard wires, plugs, and transducers commonly used by the aviation industry could function at our extreme temperatures. There were no hydraulics or pumps, oils or greases that could take our kind of heat. There were no escape parachutes, drag chutes, rocket-eject propellants, or other safety equipment that could withstand our temperature ranges, and no engine fuel available for safe operation at such high temperatures. There was no obvious way to avoid camera lens distortions from fuselage heat flows, and no existing pilot life-support systems that could cope with such a hostile, dangerous environment. We would even be forced to manufacture our own titanium screws and rivets. By the time the project ended, we had manufactured on our own thirteen million separate parts.

Cannibalization had been a house specialty at the Skunk

Works on every airplane we had ever built before this one. To save cost and avoid delays, whenever possible we would use engines, avionics, and flight controls from other aircraft and cleverly modify them to fit ours. But now we would even have to reinvent the wheel — literally. Our fear was that the rubber tires and folded landing gears might explode as the heat built in flight. We took our problem to B. F. Goodrich, which developed a special rubber mixed with aluminum particles that gave our wheels a distinctive silver color and provided radiant cooling. The wheels were filled with nitrogen, which was less explosive than air.

The airplane was essentially a flying fuel tank carrying 85,000 pounds of fuel — more than 13,000 gallons — in five noninsulated fuselage and wing tanks that would heat up during supersonic flight to about 350 degrees; we turned to Shell to develop a special, safe, high-flash-point fuel that would not vaporize or blow up under tremendous heat and pressure. A lighted match dropped on a spill would not set it ablaze. The fuel remained stable at enormous temperature ranges: the minus 90 degrees experienced when a KC-135 tanker pumped fuel into the Blackbird at 35,000 feet, and the 350 degrees by the time the fuel fed the engines. As an added safety precaution, nitrogen was added to the fuel tanks to pressurize them and prevent an explosive vapor ignition.

The fuel acted as an internal coolant. All the heat built up inside the aircraft was transferred to the fuel by heat exchangers. We designed a smart valve — a special valve that could sense temperature changes — to supply only the hottest fuel to the engines and keep the cooler fuel to cool the retracted landing gear and the avionics.

One day Kelly Johnson came to me looking as happy as a little kid who had just received a free World Series ticket. "I found a guy in Texas who claims to have developed a special oil product that can withstand nine hundred degrees," he said. "He's sending a sample overnight."

Poor Kelly. A big canvas sack of crystal powder arrived the next day. The powder changed into a lubricant at 900 degrees. Oiling our engines with a blowtorch just wouldn't make it for us, so we turned to Penn State's excellent petroleum research department to develop a special oil, which they eventually did, but at a price that made it imperative that not one drop be wasted. A quart of our oil was more expensive than the best scotch malt whiskey. We use 10–40 motor oil in our cars when wide temperature ranges are anticipated; our oil was more like 10–400.

Slowly, but expensively, we began to problem-solve. Kelly offered a hundred-dollar reward for any idea that saved us ten pounds of weight. No one collected. He offered five hundred bucks to anyone who could come up with an effective high-temperature fuel-tank sealant. No one collected that dough either, and our airplane would sit on the tarmac leaking fuel from every pore. But fortunately the tanks sealed themselves in flight from the heat generated by supersonic speeds.

Our crown of thorns was designing and building the powerful engine's inlets — the key to the engine's thrust and its ability to reach blistering speeds. This became the single most complex and vexing engineering problem of the entire project. Our engines needed tremendous volumes of air at very high pressures to be efficient, so Dave Campbell and I invented movable cones that controlled the velocity and pressure of the air as it entered the engines. These spike-shaped cones acted as an air throttle and actually produced 70 percent of the airplane's total thrust. Getting those cones to function properly took about twenty of the best years off my life.

I had a staff of three (by Skunk Works standards that was almost an empire). On the air-conditioning team, I had two engineers to help design the internal cooling system to safeguard the camera bays and the avionics and landing gear systems. The cockpit environment also presented a unique problem: without effective and fail-safe cooling the pilot could bake a cake in his lap. And as head thermodynamicist, that problem fell in *my* lap.

We designed the cockpit air-conditioning to bleed air off the engine compressor and dump it through a fuel air cooler, then through an expansion turbine, into the cabin at a frigid minus 40 degrees F, which lowered the ovenlike 200-degree cockpit to a balmy Southern California beach day. Developing these systems took us a year of frustrating trial and error.

Our engines were the only items off the shelf, so to speak. Kelly agreed with me that if we started from scratch to invent our engines, we would be hopelessly late in delivering the first Blackbird. We chose two Pratt & Whitney J-58 afterburning bypass turbojets, designed in 1956 for a Navy Mach 2 fighter-interceptor that had been canceled before the start of production. But the engine, which would need major modifications for our purposes, had already undergone about seven hundred hours of testing before the government cut off its funding. Each of these engines was Godzilla, producing the total output of the *Queen Mary's* four huge turbines, which churned out 160,000 shaft horsepower. Using afterburners at Mach 3, the exhaust-gas temperatures would reach an incredible 3,400 degrees.

This propulsion system would not only be the most powerful air-breathing engine ever devised but also the first ever to fly continuously on its afterburners, using about eight thousand gallons of fuel an hour. To build this system to our needs and specifications, P & W's chief designer, Bill Brown, who had worked closely with us on the U-2, agreed to construct a separate plant at their Florida manufacturing complex exclusively for developing this extraordinary engine. The CIA unhappily swallowed the enormous development costs of $600 million. Brown preached teamwork and pledged an unprecedented degree of partnership with the Skunk Works in general, and with me and my team in particular, to design their compressor to match my airflow inlet. This close partnership between the engine builder and the airplane manufacturer was unusual in an industry where the engine people and the airplane manufacturers often used each other as scapegoats if an airplane failed to live up to its potential. Abandoning this kind of adversarial posturing led to achieving the most powerful engine system coupled to the highest-performance inlets at these high Mach numbers that has ever been attained.

Bill Brown also offered us access to one of the largest and costliest computer systems of the day, the IBM 710. The system was state-of-the-art for its time and about as sophisticated as today's commonly used handheld calculators. But, like us, the Pratt & Whitney team would problem-solve mostly by what Kelly jokingly referred to as "my Michigan computer" — the battered old slide rule he had been using since his university days at Michigan.

Despite the unprecedented power of those two massive engines, they supplied only 25 percent of the Blackbird's thrust at Mach 3, a fact Bill Brown hated to admit. The inlets produced most of the propulsive thrust by supplying the air required by the engine at the highest pressure recovery and with the lowest drag. At supersonic cruising speed, each of our two inlets swallowed 100,000 cubic feet of air per second — the equivalent of two million people inhaling in unison. Hydrocarbon fuels like kerosene burn at high pressure, but at 80,000 feet, the air density is only one-sixteenth the density at sea level, so we used the inlets to pump up compression, before burning the air-fuel mixture inside the engine and then expanding it through a turbine and finally refiring it with tremendous thrust through the afterburner.

The only way to get energy out of the air is to pump pressure into it or to burn it. Our unique movable inlet cone, shaped like a spike, acted as an air throttle by regulating the airflow into the inlet across the spectrum of speeds from takeoff to climb to maximum cruise speed. Operated by our revolutionary electronic measuring sensors, which recorded speed and angle of attack to position the spikes precisely, the movable spikes were fully extended about eight feet out from the inlets on takeoff and gradually retracted by as much as two feet into the inlet interior as the airplane gained maximum supersonic speed.

At 80,000 feet, the outside air temperature was about minus 65 degrees F. As the inlet sucked in the air at Mach 3 through narrowed openings that compressed it, the air heated to 800 degrees. The bypass turbojet engines took the heated and high-pressure air (40 psi) and squeezed it further in a compressor, heating it to about 1,400 degrees F. At that point fuel was added to heat the air inside the burner to 2,300 degrees F. This supercharged air was then expanded through the turbine, before being fed into the roaring afterburners, superheating the combustible mix of gas and air to 3,400 degrees F, just 200 degrees below the maximum temperature for burning hydrocarbon fuels. The white-hot steel nozzle spit out its fiery plume in the form of diamond-shaped supersonic shock waves. Even in the frigid upper atmosphere, the air boiled at 200 degrees F for a thousand yards behind those booming engines. This unprecedented propulsive power sped the Blackbird at an unbelievable two-thirds of a mile a second.

About six months into our wind tunnel testing, I went to Kelly with joyful results: the inlets produced 64 percent of the airplane's full-throttled power. The precise shaping of the inlets and our unique movable air throttle cones, or spikes, allowed us to achieve an astounding 84 percent propulsion efficiency at Mach 3, which was 20 percent more than that of any other supersonic propulsion system ever built.

Developing this air-inlet control system was the most exhausting, difficult, and nerve-racking work of my professional life. The design phase took more than a year. I borrowed a few people from the main plant, but my little team and I did most of the work. In fact the entire Skunk Works design group for the Blackbird totaled seventy-five, which was amazing. Nowadays, there would be more than twice that number just pushing papers around on any typical aerospace project.

Having today's high-speed computers would have accelerated the design process and simplified much of our testing, but perfection was seldom a Skunk Works goal. If we were off in our calculations by a pound or a degree, it didn't particularly concern us. We aimed to achieve a Chevrolet's functional reliability rather than a Mercedes's supposed perfection. Eighty percent efficiency would get the job done, so why strain resources and bust deadlines to achieve that extra 20 percent, which would cost as much as 50 percent more in overtime and delays and have little real impact on the overall performance of the aircraft itself?

As it happened, we achieved 70 percent efficiency within the first half year of our work, but to tweak it above that to our target of 80 percent took an additional fourteen months. Of primary concern was where to precisely locate the supersonic shock wave within the inlet walls. That was the key to achieving maximum efficiency, because the shock wave in the wrong place in the inlet would block incoming air, causing energy loss, drag, and in a worst-case scenario — stall.

I logged hundreds of hours testing inlet shapes and cone models at NASA's Ames Research Center at Moffett Field in Northern California, a giant complex of high-speed wind tunnels. That became my second home, where I spent weeks at a time using their largest, most-powerful supersonic wind tunnel, a twenty-foot-long, ten-foot by ten-foot rectangular chamber powered by a gigantic compressor capable of driving an ocean liner, and a three-story cooling tower holding tens of thousands of gallons of water. Running Mach 3 pressures for several hours at a time drained so much electricity needed by local industry that we were forced to test only late at night, working usually until dawn. Wind tunnel tests cost us $10,000

to $15,000 an hour and we ran up a stupendous bill because
our models were tested from every angle and on more than
250,000 separate measuring points, across a broad range of
Mach numbers and pressures.

But Kelly preached that a precise model, even one like ours
that was one-eighth the size of the real inlet, would provide
precise measurements for the full-size model as well. So our
wind tunnel testing was critical to the airplane's success and
usually ended at sunrise, when our exhausted little group of
analysts finished computing the previous night's test results.
Nowadays, such calculations can be performed in a mini-
second by supercomputers.

Kelly was now so desperate to save weight that he upped
the ante to one hundred and fifty bucks to anyone who could
save him a measly ten pounds. I suggested we inflate the
Blackbird's tires with helium and give each pilot a preflight
enema. Kelly tried the helium idea, but helium bled right
through the tires. The enema idea he left to me to try to pro-
mote among the pilots.

One bet easy to collect was that we would never have this
airplane flying on time. By the end of 1960, we were over budget by 30 percent and Kelly was forced to concede we would
be at least a year late in getting the Blackbird into the sky.

The biggest delay was in my bailiwick. We had contracted
the inlet control mechanism that would move and position
the cones to Hamilton Standard, which was also doing the
fuel control system for Pratt & Whitney. The trouble was that
the pneumatic inlet controls they devised were not responding
quickly enough. We had spent $18 million to develop this sys-
tem, but after more than a year the problem was unsolved.
Finally, I took the matter to Kelly. I said, "Kelly, I think we've
got to cut our losses and find someone else to get the job
done."

He cussed and agreed. We went to a company called Air
Research and they developed an electronic control that saved

our bacon. And they did it in less than a year. Meanwhile Pratt & Whitney was struggling with a slew of problems that were putting them further and further behind schedule. Dick Bissell and his assistant, John Parangosky, watched in anguish as our delays and costs mounted. In pique, Parangosky had begun referring to the P & W engine as the "Macy's engine," and complained to Bill Brown, their program manager, rather unfairly, "If we gave as much money to R. H. Macy's, *they* could build that engine in time for Christmas." By mid-1960, the agency decided to crack a mean whip. Kelly was called on the carpet and told he had to accept a CIA engineer birddogging in his shop and looking over all shoulders. Kelly blew up. He knew the agency was only trying to cover their own asses. But Parangosky, whom Kelly liked and trusted, flew in from Washington and warned him that if he refused the request there was a grave likelihood that the CIA would cancel the contract entirely.

Kelly fumed. "No, John! I'm not gonna have one of your spies poking into my business. Bissell promised me you guys would keep hands off and let me do this thing my way just like the U-2."

"Kelly, be reasonable. We won't get in your way. We just want someone here you can trust and we can too."

He suggested a very bright and able engineer on the CIA's payroll named Norm Nelson, whom Kelly had known from World War II days, and both liked and respected. Kelly sulked but ultimately surrendered. "Well," he said, "I'll let in Norm Nelson, but not another goddam person. You got that? Besides, you tell Nelson he can have a desk, a phone, but no chair. I expect him in the shop, not sitting on his fat duff."

Nelson, who arrived in the spring of 1960, became the first outsider ever allowed a place inside Kelly's realm. He gave Norm a free hand and actually took suggestions because he respected Norm's judgment. We all knew that he reported directly to Bissell, and he knew that we kept him out of certain meetings and didn't let him in on *everything*. But Norm was sharp. I recall one meeting in 1961, when Kelly told Norm that the agency had given us an additional $20 million to develop wing tanks on the Blackbird to extend its range. Norm did some quick calculations and figured out we would extend the airplane's range only eighty miles. "Fooling around with wing tanks at this point will be more trouble than it's worth," Norm insisted. Kelly said, "You're probably right." Kelly Johnson sent back the $20 million that afternoon.

But our biggest problem was about as easy to conceal from Norm as a pregnant pachyderm on top of a flagpole. Norm Nelson came aboard just as we were starting to build a mock-up of the fuselage-cockpit section, which would contain more than six thousand parts, for heat testing inside an oven. To our horror, we discovered that the titanium we were trying to use was as brittle as glass. When one of the workers dropped a piece off his bench, it shattered in a dozen pieces. The trouble was diagnosed as poor quality-control procedures in the man-ufacturer's heat-treatment process — a problem that caused endless delays, forcing us to reject 95 percent of the titanium delivered and set up a rigorous quality-control procedure.

For an outfit that detested red tape, we now found ourselves wallowing in bureaucratic procedures. We sample-tested for brittleness three out of every ten batches of titanium received and kept detailed records of millions of individual titanium parts. We could trace each part back to the original mill pour, so if a part went bad later on, we could immediately replace other parts from that same batch before trouble developed.

We also learned the hard way that titanium was totally in-compatible with many other elements, including chlorine, fluorine, and cadmium. When one of our engineers drew a line on a sheet of titanium with a Pentel pen, he discovered that the chlorine-based ink etched through the titanium just like acid. Our mechanics working on the engine installation used cadmium-plated wrenches to tighten bolt heads. When the bolts became hot, the bolt heads just dropped off! It took intensive detective work to zero in on the cadmium contamination culprit, and we quickly removed all cadmium-plated tools from toolboxes. Even the routine matter of drilling a hole became a nagging frustration. When machining standard aluminum, a hundred holes could be drilled without resharpening the bit. With titanium, we had to resharpen every few minutes and were forced to develop special drills using special cutting angles and special lubricants, until we finally were able to drill more than 120 holes before having to resharpen. But it took us months of painstaking experimentation to get that far.

Miles of extrusions were required to produce an aircraft the size of the Blackbird, so it was necessary to invest time and more than a million dollars in new state-of-the-art precision drills, cutting machinery, powerheads, and lubricants.

For each problem solved, two or three others suddenly cropped up. We were stunned when spot welds on panels began to fail within six or seven weeks. Some intensive sleuthing revealed that the panels had been welded during July and August, when the Burbank water system was heavily chlorinated to prevent algae growth. The panels had to be washed after acid treatment, so we immediately began using only distilled water. During heat tests, the wing panels warped so badly they looked like potato chips. We worked for months to find a solution and finally used corrugated panels that allowed the metal to expand without warping under immense heat friction. At one point an exasperated Kelly Johnson told me: "This goddam titanium is causing premature aging. I'm not talking about on parts. I'm talking about on *me*."

We set up training classes for machinists using titanium for the first time and a research operation for developing special tools that would make their jobs easier. Between the new machines and the training, our bean counters figured that ultimately we saved $19 million on the production program.

Still, unforeseen problems kept increasing the costs, and only a few days after the November 1960 elections, which brought John F. Kennedy into office and took the Republicans out, Kelly returned from a week's vacation and found a wire awaiting him from Dick Bissell, inquiring about what it would cost the government to cancel the Blackbird program. "I am very afraid," Kelly noted in his private log, "about what will be Kennedy's attitude toward the program, its overall cost increase, which is very high on all fronts, and the fact that our Russian friends have now come up with a new Tall King radar which appears to be capable of detecting a target about one-third the size that we are able to accomplish with the Blackbird. With all this we have made remarkable strides in reducing the radar cross section, and our experts say we would have about one chance in 100 of being detected, with practically no chance of being tracked."

Our chief chemist, Mel George, helped us to develop special antiradar coatings loaded with iron ferrites and laced with asbestos (long before it became a dirty word) to be able to withstand the searing heat from the tremendous friction hitting the leading edges of the airplane. These coatings were effective in lowering the radar cross section and comprised about 18 percent of the airplane's materials. In effect, the Blackbird became the first stealth airplane; its radar cross section was significantly lower than the numbers the B-1B bomber was able to achieve more than twenty-five years later.

To save time and money and maintain high quality standards, we did our own milling and forging and at one time approached the ability of our vendor's plants to roll parts to precise dimensions. We even developed our own cutting fluid that would not corrode titanium. To prevent oxidation of the titanium — which caused brittleness — we welded in specially constructed chambers with an inert nitrogen gas environment. In all we had about twenty-four hundred trained fabricators, machinists, and mechanics working on the project, all of them specially trained and carefully supervised. And at the height of production, in the mid-1960s, we employed a huge force of nearly eight thousand workers and delivered one Blackbird per month. While we were trying to build that first airplane, the unions were giving Kelly fits because he ignored seniority rules and chose the best workers, so Kelly had the union heads cleared and walked them through the plant and showed them the airplane. He said, "Gents, this airplane is vital for our nation's security. The president of the United States is counting on it. Please don't get in my way here." They backed off.

For security and other reasons, the airplane was assembled in various buildings in the complex. One unique, extremely time-saving technique was to build the fuselage on the half shell. The left half and right half were assembled independently to create easier worker access, then fit together and riveted into place. That was a major first in aircraft manufacturing.

Other Voices
Keith Beswick

I began working for the Skunk Works in flight-test operations on the U-2 out at Edwards Air Force Base in October 1958. By the 1960s I was put in charge of flight testing for the Blackbirds. We were working on the cutting edge, forced to improvise a dozen times a day. We would rig up some of the damndest tests ever seen. I remember when Ben Rich and his cohorts decided to test their cockpit air-conditioning system, they put one of our test pilots inside a broiler big enough to roast an ox medium rare, to see if their cooling system really worked well enough. The guy sat inside a cylinder cooled to 75 degrees by Ben's air-conditioning system while the outer skin of the cylinder cooked to about 600 degrees. I asked Ben,

"What would you do if the system failed?" He laughed. "Get out of town in a hurry."

During the test phase of the Blackbird, we pumped air pressure into the fuel tanks up to one and a half times greater than the design limits. We did this late at night, inside Building 82, when there were very few people around, because if you're pumping up that much titanium and if there should be a major failure and the thing blows — that's an awful lot of energy bursting like a balloon. It would blow out windows in downtown Burbank, so we filled the fuselage with several million Ping-Pong balls to dampen any explosive impact and hid behind a thick steel shield with a heavy glass window, watching the airplane getting all this high-pressure air pumped into its tanks. We were pumping up to twelve inches of mercury and got to about ten when suddenly, Kaboom! The drag chute compartment in the rear blew out. Henry Combs, our structural engineer, took a look at the damage and went back to the drawing board and made the fixes. A few nights later we were back behind the protective shield in Building 82. This time we got up to ten and a half inches of mercury when the drag chute forward bulkhead ruptured with a loud bang. Henry took notes and went back to the drawing board. Three nights later we were all back for more testing. The pumping began and we heard the airplane crickling and crackling as the pressure mounted. It was really tense behind that shield as the mercury rose. We got up to eleven and a half inches of mercury and heard the airplane go crick, crack, crick. And Henry shouted, "Okay, stop. That's close enough."

In January 1962 we were ready to cart the Blackbird out to the test site. The airplane was disassembled into large pieces and would be trucked out in a heavily guarded wideload trailer, 105 feet long and 35 feet wide. Dorsey Kammerer, head of the flight-test shop at that time, came up with the idea of driving the entire route ahead of time using a pickup truck with two bamboo poles up on top. One pole was as wide as the load would be going along the edge of the series of freeways and underpasses. The second pole was exactly as high as the load. They drove the entire route, and any traffic or speed signs that hit against the pole, they pulled over and used a hacksaw to cut the sign off. Then they fit the pieces back with a brace and bolt and marked the sign. On the day we moved the airplane under wraps the lead security car stopped at all the marked signs, undid the bolts to take down the sign while the truck passed, then the rear security car bolted the signs back in place and the convoy moved on. But not even that kind of efficiency could overcome the unexpected disaster. Midway into the trip, a Greyhound bus passed us too closely and was scraped. Our security guys flagged him over, haggled for a while with the driver, and paid him $3,500 cash in damages right on the spot — to keep any official insurance or accident report from being filed involving the most top secret truck caravan in America.

We were scheduled to fly the airplane for the first time only thirteen days after we got it out to the test site. The J-58 engines weren't ready, however, but Kelly didn't want to wait, so in typical Skunk Works fashion, we reengineered the insides of the engine mounts to put in lesser-powered J-75s. The fuel, JP-7, has a kerosene base and such an extremely high flash point that the only way to ignite it was by using a chemical additive called tetraethyl borane, injected during the start procedure.

The first time we tried to test the engines, nothing happened. They wouldn't start. So we rigged up two big 425-cubic-inch Buick Wildcat race car engines, an estimated 500 horsepower each, to turn the massive starter shafts and those suckers did the trick. The hangar sounded like the damned stock car races, but starting those huge engines was tough. The engine oil, formulated for high temperatures, was practically a solid at temperatures below 86 degrees. Before each flight, the

oil had to be heated and it took an hour to heat it 10 degrees.
But once those engines roared to life, it was a sight to behold.
Twenty seconds into takeoff, the Blackbird achieved 200 mph
in forward speed.

Every time I saw that Blackbird on a runway I got goosebumps. It was the epitome of grace and power, the most beautiful flying machine I've ever seen. I was up in the control tower
for the April 25th high-speed taxi test. Our test pilot, Lou
Schalk, headed down the runway and over-rotated the engines
slightly so that the airplane became airborne for a few seconds,
wobbling back and forth. I thought Lou would stay airborne
and circle around and land, but instead he put it back down
right then and there in a big cloud of dust on the lake bed. For
a moment, my heart stopped. I couldn't tell whether or not he
crashed. And it seemed an eternity before the nose of the airplane appeared out of a cloud of dust and dirt, and I heard
Kelly's angry voice over the radio, "What in hell, Lou?"

10

GETTING OFF THE GROUND

THE BLACKBIRD was a wild stallion of an airplane. Everything about it was daunting and hard to tame — building it, flying it, selling it. It was an airplane so advanced and awesome that it easily intimidated anyone who dared to come close. Those cleared to see the airplane roar into the sky would remember it as an experience both exhilarating and terrifying as the world shook loose. Richard Helms, then a high-level CIA executive, recalled watching a Blackbird take off on a night flight from our secret base in the 1960s, with the roar of an oncoming tornado and the ground shaking under his feet like an eight-point earthquake, as the engines spouted blinding diamond-shaped shock waves. "I was so shaken," Helms told me recently, "that I invented my own name for the Blackbird. I called it the Hammers of Hell."

A few months after the first successful Blackbird test flight in April 1962, test pilot Bill Park appeared at my desk and dropped his plastic flight helmet in my lap. "Goddam it, Ben, take a look at that," he said, pointing to a deep dent near the crown. As Bill described it, he was cruising at sixty-five thousand feet on a clear, crisp morning above New Mexico, when suddenly, with his airplane blistering at 2.7 Mach, he was deafened by a loud bang and violently flung forward in his harness, smashing his head against the cockpit glass and almost knocked unconscious. "It felt like a couple of the L.A. Rams shaking me as hard as they could," Bill said. The problem was called an "unstart." It occurred when air entering one of the two engines was impeded by the angle of the airplane's pitch or yaw and in only milliseconds decreased its efficiency from 80 percent to 20 percent. The movable-spike inlet control could correct the problem in about ten seconds, but meanwhile the pilot was flung around helplessly, battered all over the cockpit. Bill Park and Lou Schalk and several of our other pilots were experiencing these awful "unstarts" as much as twenty times in ten minutes. The damndest part was that the pilot often couldn't tell which engine was affected and sometimes he turned off the wrong one to get a relight and was left with no power at all. This happened to a Blackbird over West Virginia. The pilot struggled to relight both engines as the airplane plunged toward earth. Finally at thirty thousand feet, the two engines came alive with a tremendous sonic boom that shattered windows for miles and toppled a factory's tall chimney, crushing two workers to death.

"Fix it!" the pilots demanded. Easier said than done, I discovered. In spite of my best efforts I never really solved the unstart problem per se. The best I could do was invent an electronic control that was basically a sympathetic unstart. If one engine was hit with an unstart, this control ensured that the second engine dropped its power too, then relit both engines automatically. In the cockpit, the pilot would be spared a near heart attack by a loud bang followed by a series of severe jolts that marked an unstart. When the new system functioned he would not even be aware that an unstart had occurred. But before I could solve the problem, I took a lot of heat. Bill Park insisted that I get off my duff and see firsthand what he and the rest of our test crew were going through. Kelly, with a diabolical glint in his cold eyes, eagerly agreed.

"Rich, goddam it, suit up and get out there and fix that goddam thing before one of our pilots breaks his goddam neck," the boss decreed. In a crazy moment of weakness, I actually agreed to fly. I got as far as the high-compression chamber, which simulated ejection, as part of my preflight briefing. In order to fly at ninety thousand feet I had to be checked out in the pressure suit in an altitude chamber, in case we lost cabin pressure or I was forced to eject in an emergency. The chances that I would experience such calamities were near zero since I would have already dropped dead from fright. Nevertheless, I found myself inside a heavy helmet and pressure suit, and the minute that chamber door slammed shut I experienced an immediate claustrophobic panic. I was sucking oxygen like a marathon runner and screaming, "Get me out of here!" Call me a coward. Call me hopeless. Call me a taxi. I bugged out.

Other Voices
Norman Nelson

I was the CIA's engineer inside the Skunk Works, the only government guy there, and Kelly gave me the run of the place. Kelly ran the Skunk Works as if it was his own aircraft company. He took no crap and did things his own way. None of this pyramid bullshit. He built up the best engineering organization in the world. Kelly's rule was never put an engineer more than fifty feet from the assembly area. But the payoff came watching that Blackbird take off on sixty-four thousand pounds of thrust blasting out from those two giant engines. We all knew it was the greatest airplane ever built and it carried the world's greatest cameras. From ninety thousand feet — sixteen miles up — you could clearly see the stripes on a parking lot. Baby, that's resolution! The main camera was five feet high. The strip camera was continuous, and the framing camera took one picture at a time. Both took perfect pictures while zipping past at Mach 3. An unbelievable technical achievement. The window shielding the cameras was double quartz and one of the hardest problems confronting us. We also had awful reflection problems and heat problems, you name it.

Because of the tremendous speed and sonic boom, we were

very limited to where we could overfly the United States during training missions. We had to pick the least-populated routes. After President Johnson's public announcement about the airplane in the fall of 1964, Kelly began receiving all kinds of complaints and threats of lawsuits from communities claiming the Blackbird had shattered windows for miles around. A few times we announced a bogus flight plan and then sat back and watched the phony complaints pour in. But some complaints were for real. One of the guys boomed Kelly's ranch in Santa Barbara as a joke that backfired because he knocked out Kelly's picture window. Another of our pilots got in engine trouble over Utah and flamed out. The Blackbird had as much gliding capacity as a manhole cover, and it came barreling in over Salt Lake, just as our pilot got a restart and hit those afterburners right above the Mormon Tabernacle. There was hell to pay.

We had to clear FAA controllers along the flight paths, otherwise they'd think they were seeing flying saucers at Mach 3 plus on their radar screens. In the amount of time it took to sneeze, a pilot flew the length of ten football fields.

We couldn't overfly dams, bridges, Indian ruins, or big cities. We had to clear and train the tanker crews of the KC-135s that carried our special fuel. Air-to-air refuelings were very tricky because the tanker had to go as fast as it could while the Blackbird was throttled way back, practically stalling out while it filled its tanks. During a typical three-to-five-hour training exercise, our pilot might witness two or three sunrises, depending on the time of day.

Another weird thing was that after a flight the windshields often were pitted with tiny black dots, like burn specks. We couldn't figure out what in hell it was. We had the specks lab tested, and they turned out to be organic material — insects that had been injected into the stratosphere and were circling in orbit around the earth with dust and debris at seventy-five thousand feet in the jet stream. How in hell did they get lifted

up there? We finally figured it out: they were hoisted aloft from
the atomic test explosions in Russia and China.

That airplane pushed all of us to our limits in dealing with
it. A pilot had to have tremendous self-confidence just to set
foot inside the cockpit knowing he was about to fly two and a
half times faster than he ever had before. I know that Kelly was
determined to spread the Blackbird technology onto the blue-
suiters and make the whole damned Air Force sit up and pay
attention to what he had produced. But I never gave him much
chance to sell a lot of these airplanes because they were so far
ahead of anything else flying that few commanders would feel
comfortable leading a Blackbird wing or squadron. I mean this
was a twenty-first-century performer delivered in the early
1960s. No one in the Pentagon would know what to do with it.
That made it a damned tough sell even for Kelly.

Kelly was his own salesman. He traveled to Washington pitching those in high places at the Pentagon and on the Hill. He
was plugged in at the CIA and knew what the top Pentagon
brass were worrying most about at any given moment, and in
the early months of the Kennedy administration many officials were popping Valium.

There were storm warnings flapping over Red Square. The
Russians seemed eager to severely test our new young president and backed up that belligerency with worrisome crash
weapons projects. The CIA had intercepted Russian telemetry
data on what they thought was a missile test in Soviet Siberia
in the spring of 1961. They sent this data to the Skunk Works
for our analysis and verification. Our telemetry experts reported back a chilling contradiction: that was no missile being
tested but a prototype supersonic bomber, the so-called Backfire, rumored to have been in the works for several years. We
were likely looking at an aircraft capable of sustained Mach 2
speeds, flying at sixty thousand feet and with an impressive

range of three thousand miles. If we were right, this was a major upset of the then current military balances of power: the Soviets were building a bomber that could come and get us, and the brass at the Air Defense Command in Colorado Springs could only look up and shake a fist because we had nothing flying that could intercept it, or any missile to shoot it down.

But fitted on the jigs of Assembly Building 82 was the frame of our Mach 3 Blackbird, being built as a CIA spy plane, which could be adapted as a high-performance interceptor that would stop the Russian bombers long before they could reach any American targets. Once the U.S. early-warning radar net (so powerful, it could track a baseball-size object from five thousand miles away) picked up a Soviet bomber force streaking toward North America, our Blackbirds could race to meet and intercept them over the Arctic Circle, beyond range of their nuclear-tipped missiles targeted against U.S. cities.

That was Kelly Johnson's Pentagon sales pitch in the dawning of the Kennedy years: the Blackbird was exactly what the brass of the fighter command should have been looking for, but, unfortunately, our airplane was so secret and knowledge of this project so limited that very few Air Force commanders knew of its existence and Kelly could not pitch anyone who wasn't on a very select list of those cleared to know about this top secret airplane. He was so constrained by security that he was practically talking to himself.

Among a few, highly placed Air Force brass who did know about our airplane there were mixed feelings about Blackbird's $23 million cost (the technology was not bargain bin) because a general would always prefer commanding a large fleet of conventional fighters or bombers that provides high visibility and glory. By contrast, buying into Blackbird would mean deep secrecy, small numbers, and no limelight. In the military, less was definitely not more. Most military officers were assigned commands or Pentagon desk jobs for three to five years, before moving on. The future uses of a revolutionary airplane like the Blackbird as a fighter or bomber was a question they would gladly leave for their successors to mull over; they aimed to make their mark quickly by putting as much new rubber at the ramp as soon as possible and earn commendations and promotion up the chain of command. Kelly Johnson's technological triumphs were thrilling to hear about but not immediately advantageous to an ambitious colonel lusting for his first star.

Kelly knew what he was up against, but he tried to improve the odds by producing the kinds of "add-ons" that no blue-suit customer could resist. For example, he put me in charge of a feasibility study for using the Blackbird as a platform for launching ICBM missiles. Launched from, say, sixty thousand feet, a missile could travel six to eight thousand miles by eliminating the tremendous fuel consumption of a ground-based launch. We even dreamed up the creation of an energy bomb that used no explosive device. Flying at Mach 3 and eighty-five thousand feet, we'd drop a two-thousand-pound weight of high-penetrating steel that would hit the ground with the force of a meteor — at about one million foot-pounds of energy and blast a hole 130 feet deep. The Air Force was interested, but fretted about the absence of a guidance system to assure pinpoint accuracy and resisted our suggestions to try to develop such a system. To the new secretary of defense, Robert McNamara, an energy bomb was futuristic drivel. McNamara had enough to worry about in the present tense.

Soviet military operations were at a higher level of alert than we had seen in years. Increasing numbers of long-range transcontinental flights by Soviet Bear bombers, capable of dropping nuclear weapons, were being made nonstop from bases in southern Russia to Havana, Cuba. The Bear was their version of our B-52. The journey was about six thousand miles, with probably two air-to-air refuelings. The Soviets insisted that these were merely long-duration training exercises. But the gut-churning fact was that if the Soviet Bears could reach Havana, they sure as hell could reach New York or Washington or Chicago. That message was like a brick hurled through McNamara's office window.

An even bigger worry was Soviet submarines with Polaris-type nuclear-tipped missiles on board that were brazenly operating in international waters off our major cities on both coasts. We figured that their primary target would be our SAC bases, which was why none of our bomber wings were located east of the Mississippi but were scattered throughout the Great Plains in central states like Nebraska and Montana. This location gave SAC a few extra minutes to get our B-52s safely into the air before enemy missiles hit. Every time a Soviet sub was spotted off our coast, our entire U.S.-based bomber fleet went on alert.

The first time Kelly met McNamara he found him haughty and cold. "That guy will never buy into a project that he hasn't thought up himself," Kelly remarked at a staff meeting soon after. "He's petty, the kind who will throw out any project begun under Eisenhower. He just doesn't believe that anyone else has his brains and he'd love to stick it to an old-timer like me just to show the entire aerospace industry who's boss."

Our contract to build the new Blackbird spy plane for the CIA was rock solid, even though our original budget estimate was now almost doubled by delays, expensive materials, and technical problems to $161 million. To compensate for these increased costs the agency had scaled back its original purchase order from twelve airplanes to ten. So it seemed unlikely that McNamara, nicknamed Mac the Knife in the corridors of the Pentagon for his slashing budget cuts, would want to put more money into a Blackbird supersonic bomber. The Air Force was already spending millions developing the North American B-70, a huge triangular-shaped monster, capable of Mach 2 speeds. The B-70 was the favorite project of

Kennedy's gruff Air Force chief of staff, General Curtis LeMay, who usually got his way simply because few civilian officials (or uniformed generals for that matter) found the courage to try to face him down. Kelly was one of the rare exceptions. He told LeMay flat out that from what he had seen of the plans the B-70 would be obsolete before it was even off the drawing board. LeMay was furious, but a lot of blue-suiters privately agreed with Kelly. The B-70 had six engines to the Blackbird's two. Our airplane was nearly twice as fast, but LeMay told Kelly he didn't know beans about bombers, to stick to spy planes and mind his own business.

But then Dick Bissell got into the act. Bissell briefed President Kennedy on the CIA Blackbird project and told him the spy version of the airplane would be operational in less than a year. When he learned how fast and how high it would fly, the new president was astonished. He asked Bissell, "Could Kelly Johnson convert your spy plane into a long-range bomber?"

Bissell replied that Kelly aimed to do precisely that. "Then why are we going ahead with the B-70 program?" Kennedy asked. Bissell shrugged. "Sir," he replied, "that's a question more properly addressed to General LeMay."

The president nodded sheepishly. But Kelly was embarrassed by Bissell's indiscretion. As he noted in his private journal, "Bissell recounted his conversation about a bomber version of the Blackbird with the President. It was not right. The President asked for our proposal for the bomber before the Air Force had even seen one and I felt obligated to rush to Washington and present it as quickly as possible to our Air Force friends and showed the proposal to Gen. Thomas White. Lt. Gen. Bernard Shriever was there and they were all very upset, as was Gen. LeMay, about losing B-70s to our airplane. But at least they fully understood that that was not my doing and they cannot control Dick Bissell's approach to the President."

But who would have guessed that Bissell's days were numbered? By April 1961, he was on his way out, his brilliant career shattered by the Bay of Pigs fiasco. Bissell had been overseeing the invasion attempt, staged by the agency using Cuban exiles trained in secret Florida camps. It was a botched mission from start to finish, and all of us were deeply depressed that Bissell had to fall on the sword along with his boss, Allen Dulles. Bissell was godfather to the Skunk Works. He started the U-2 and that really put us in business to stay. But he took his fall gracefully, and in April 1962, even though he was out of government by then, Kelly invited him out to the secret base to watch the first flight of the Blackbird. Both men were as tough as titanium, but both were clearly moved watching our test pilot, Lou Schalk, gun those two tremendous engines and rip into the early-morning cloudless sky. It was one of those unmatched moments when all the pain and stress involved in building that damned machine melted away in the most powerful engine roar ever heard.

Kelly worried that, with Bissell gone, Mac the Knife might convince the president to cut out the expensive Blackbird CIA operation altogether and cancel us before we had a chance to prove our worth collecting radar and electronic intercepts along the borders of the Soviet Union. From our great heights we could penetrate hundreds of miles into Russia with side-looking radar without actually crossing their borders. But in June 1961, the new president attended his first summit, in Vienna with Khrushchev, trying to de-escalate tensions with the Russians over the future status of Berlin. The meeting with the Russian leader was so unnervingly hostile that JFK came away privately convinced that we and the Russians were on the brink of war.

At the Skunk Works we sensed those mounting tensions immediately, when Air Force Chief of Staff LeMay made his first trip to Burbank to see the Skunk Works for himself. The trip itself was highly unusual because Kelly and Curtis were not exactly buddies and had pretty well avoided each other. The word we got was that Curtis LeMay wasn't paying a courtesy call, but was bringing his checkbook.

Still, Kelly was wary. LeMay blamed Kelly for the administration's decision to suddenly cut back on the B-70 program from ten bombers to only four. He thought Kelly and Bissell had connived to sabotage his B-70 with Kennedy. LeMay was also sore about our close relationship to the CIA because in his view the agency had no right to have its own independent air wing, furnished by the Skunk Works. But now LeMay whisked in with his entourage and a shopping list that included converting the Blackbird into an extended-range deep-penetration bomber that the Russians could not stop. LeMay had fathered the SAC strategy called MAD — Mutual Assured Destruction. That said it all. Our Blackbird would nuke 'em back to the Stone Age.

Kelly briefed LeMay personally and invited several of us who were experts on various components of the airplane to sit in just in case the general had technical questions. I was intrigued watching the big, two-fisted LeMay puffing on a thick cigar, his shrewd eyes focused in concentration as Kelly zipped through a classified slide show detailing all the performance characteristics of the new airplane.

"Could you fire air-to-ground missiles from that airplane while going at Mach 3 speed?" the general asked Kelly. Kelly replied affirmatively. "We've done the theoretics on this, General, and we feel confident that we could do this successfully."

"Could you guide a missile to within two hundred feet of a target?" LeMay asked. Kelly again said yes, theoretically, and added that if the missiles were nuclear, pinpoint accuracy was unnecessary. Used as a deep-penetration tactical bomber, the Russians couldn't stop us. Used as an interceptor, we were also unstoppable. Using look-down, shoot-down air-to-air missiles, our high-flying interceptor could defend all of North America against any long-range bomber force that would be expected to fly low to the deck to avoid radar detection. "With our speed," Kelly insisted, "fewer interceptors would be needed to cover the entire North American continent. We'd look down at the Soviet bomber fleet and pick them off like fish in a barrel." Kelly declared, "General, the fellow who gains the high ground takes the battle. Our speed gives us our height. We're king of the mountain."

LeMay suddenly raised his hand as a signal for Kelly to stop talking. Then he stood up, grabbed Kelly by the arm, and led him to the far corner of Kelly's huge office for a private, whispered conference that lasted nearly ten minutes, while the rest of us sat transfixed, watching these two titans of military aviation cooking up some sort of scheme or scenario.

Kelly told us later that LeMay was enthusiastic about using the Blackbird as an interceptor but resisted the idea of using it as a bomber. The B-70 was still very much on his mind. "Johnson, I want a promise out of you that you won't lobby any more against the B-70." Kelly agreed — a promise he would deeply regret in the years ahead. "We'll buy your interceptors. I don't have a number yet but I'll get back to you soon."

Kelly asked, "What about the reconnaissance aircraft we built for the agency? Can't the Air Force use any?" LeMay looked dismayed. "You mean, we haven't ordered any?" He wrote a note to himself and promised Kelly he would forward an Air Force contract for the two-seater version of the spy plane within a few weeks.

The very next day, Kelly was tipped off by a colonel on LeMay's staff that on the trip back to Washington aboard his jet, LeMay revised his thinking rather sharply and ordered his staff to develop a proposal for building ten Blackbird interceptors and ten tactical bombers a month!

For once, Kelly was speechless. "If this really comes to pass," he told me, "our whole concept of operation will have to change. This would be such a staggering operation, when you consider it means building one airplane every other working day, that I doubt we could do it successfully using our Lockheed facilities here in Southern California. I could be wrong, but let's do a study and see precisely what it would take."

The size of LeMay's shopping list reinforced our feeling that the administration was quietly and quickly gearing up for a major military showdown with the Russians. That conviction grew a few weeks later when Secretary McNamara himself arrived at our doorstep under a tight lid of secrecy, to be briefed by Kelly, review our management methods on the project, and see the Blackbird for himself. He brought with him all his top people: Secretary of the Air Force Joseph Charyk, Assistant Defense Secretary Roswell Gilpatric, and the Air Force's future secretary, Harold Brown. Kelly briefed, while McNamara took copious notes and asked several questions about the airplane's unique navigational system. It was the first astro-navigational system that actually used a small computer-driven telescope to find approximately sixty stars in its database. The telescope looked through a small window toward the rear of the airplane and was extremely accurate and reliable locking onto stars. It was so sensitive that it once locked onto a rivet hole in the hangar roof when the system was accidentally turned on during airplane servicing.

McNamara wanted to know all about it. "This is the most accurate navigational system we can devise," Kelly said. "Remember, we are traveling at twice the speed of a sixteen-inch shell, and we don't turn on a dime. A tight turn takes between sixty and a hundred nautical miles, and if a pilot gets a little sloppy he could start a turn over Atlanta and end up over Chattanooga."

Several of us escorted the official party during the inspection tour of the Blackbird inside the giant assembly building. One of the generals on the secretary's staff took me aside. "Mr. Rich," he said, "I don't understand why you're building those large spikes to block air coming into the inlet. What is the

principle here? It seems to me you'd want the air unimpeded."
I couldn't believe that an Air Force general would ask such a
naive question. I said, "General, the object is to build up pressure at high altitudes. Did you ever try to squirt water from a
hose by placing your thumb over the opening?" His nod
showed a glimmer of understanding.

By the time McNamara and his party boarded their jet
for a red-eye flight back to Washington, Kelly was practically
dancing a polka. The briefing could not have gone better.
We expected purchase orders to follow. And we weren't disappointed. As an immediate follow-up, Lew Meyer, the Air Force
assistant secretary for finance, flew out the following week
and informed us that we would probably get orders for ten
Blackbird reconnaissance aircraft for the Air Force, in addition to the ten we were building for the CIA. They wanted
a larger two-seater version, with a pilot up front and a
navigator-electronics specialist called the Reconnaissance
Systems Officer in a separate rear cockpit, working the
routing, as well as operating all the special avionics for captur-
ing enemy radar frequencies and electronic intercepts. We
would also be receiving a contract for ten fighter-interceptor
versions of the Blackbird and twenty-five tactical bombers.
These orders would be worth hundreds of millions of dollars
and seemed to defy the longtime Pentagon quasi-socialistic
policy of never putting all their purchases in one manufacturer's shopping cart.

Spreading the profits among the key manufacturers was the
usual military-industrial game, but it really seemed that the
era of the Blackbird was at hand: Blackbird spy planes, Blackbird interceptors, and Blackbird bombers. The Blackbird so
outmatched any other airplane in the world with its speed and
altitude, it would dominate air warfare for at least a decade or
more.

Before 1962 ended, Kelly had obtained approval from Lockheed management to construct a million-dollar engineering

building on the strength of the proposed expansion of the
Blackbird programs. Our in-house study, meanwhile, indicated we would need $22 million in additional funds to in-
crease production facilities to meet the large production
quotas of Mach 3 Blackbird fighters and bombers the Air
Force expected to order.

All of us began counting our big bonuses long before the
check was in the mail. And all of us were in for one big let-
down. I remember my producer brother telling me once about
a young filmmaker who had been courted by a big studio, actually being kissed on both cheeks by the reigning mogul and
declared a genius, and his script given the top priority for
quick production. But a month later, that same young genius
could not even get his calls returned. He had gone from hot to
cold with no apparent explanation. That typical Hollywood
story became the metaphor for the Blackbird scenario as
played out by the Kennedy administration. Suddenly, Kelly
could not get calls returned from key administration players
making decisions about Blackbird bombers and interceptors.
A few top generals began ducking him, too, and the word
drifting back to us from the Pentagon was that McNamara's
young Turks advising him on cost-effectiveness refused to believe that the Russians were actually developing a supersonic
Backfire bomber. Without that threat, the Blackbird was not a
necessary deterrent. McNamara's Turks insisted we could get
by using the old standby fighter, the Convair F-106. This was
the updated version of the F-102 delta-wing fighter that was
the backbone of the Air Defense Command. Both of these airplanes had only supersonic dash (Mach 1.8) capability, their
fuel capacity limited to a scant five minutes of supersonic
flight.

Kelly thought Mac the Knife had taken leave of his senses.
But he was not in the loop about planning for possible U.S.
military intervention in Vietnam and the secret preparations
for air and ground action by our forces. As a first step,

McNamara approved a new classified prototype for the first swing-wing tactical fighter, called the TFX (Tactical Fighter Experimental), that would specialize in ground attack. Built by General Dynamics, it was destined to be one of the most controversial airplanes of the era, involving tens of millions in cost overruns, as its designers struggled to solve the horrendous problems of a movable wing — parallel on takeoff and becoming swept-back as the airplane gained Mach 2 speed. The Air Force wanted the F-111, as it was officially designated, to come in on the deck, evading radar by hiding in the ground clutter, to support local troop action in hostile ground action. But the first F-111s used in Vietnam took tremendous losses because their terrain-following radar, which allowed them to skim over treetops in the dead of night, acted like a powerful beacon for enemy radar to home in on. Even worse, most of us in the Skunk Works thought that the minute look-down, shoot-down radar fire control systems were perfected, airplanes like the F-111 would become obsolete. A higher-flying MiG pilot looking down on a squadron of F-111s could become a combat ace in seconds.

To prove the Blackbird's tremendous performance capabilities to McNamara, we launched a flight in May 1962, from Edwards Air Force Base outside Los Angeles to Orlando, Florida, that blazed across the country in one hour and twenty-eight minutes. And if he wasn't paying attention, we sent another Blackbird winging east from San Diego, which arrived over Savannah Beach, Georgia, only fifty-nine minutes later. We also started development of a weapons system for the Blackbird that would demonstrate how easily an airplane like the F-111 could be shot down while flying at the treetops.

This weapon was look-down, shoot-down radar and air-to-air missiles. Developing it, we were twenty-five years ahead of anyone else. We took an existing $80 million air-to-air missile developed by Hughes, called the GAR-9, and a Westinghouse ASG-18 radar system, both created for the Navy, and augmented them with our own special fire control system. The result stunned the blue-suiters, who thought it was impossible to successfully fire a missile from an airplane speeding at three times the speed of sound. Most air-to-air missiles traveled five, ten, fifteen miles to targets locked on at the same altitude or slightly below or above. We locked onto targets more than a hundred miles away and tens of thousands of feet below. From 80,000 feet, we knocked out drones flying on the deck at 1,500 feet. From 87,000 feet, we hit a drone flying at 40,000 feet.

To keep the missile from hitting our own airplane, we added a trapeze device to the missile launch system and used ejection cartridges to drop the missile nose down before it fired and sped off to the target at Mach 6. Our first test of the new system occurred in March 1965. We hit a drone from 36 miles away at a closing rate of 2,000 mph. A few weeks later, we fired the missile from 75,000 feet, while traveling at Mach 3.2, and hit a drone flying at 40,000 feet, 38 miles away. But a few months later we really rocked the Air Force. From 75,000 feet, we hit a low-altitude remote-controlled B-47 flying over the Gulf of Mexico at 1,200 feet from 80 miles off.

Kelly was overjoyed. Our testing was a superlative twelve hits out of thirteen attempts — firing from all heights and ranges. It was the most successful new weapons test in history. Our missile should have blasted a big hole in the administration's backing of the F-111, which now was proven by our tests to be obsolete before it even flew. We knew the Russians were crashing a program to develop this kind of look-down, shoot-down system. Once it was in place aboard their newest MiG interceptors, life would be hell for any pilot in a low-flying aircraft like the F-111.

Our message to the blue-suiters: tell McNamara he's backing the wrong bomber. Their message back: tell him yourself; he won't listen to us. Kelly flew to Washington in the winter of 1966 and stormed in on Air Force Secretary Harold Brown,

who was also rather blunt, and the two went head to head.
Kelly called the F-111 a national scandal if the administration
forced it on the Air Force in spite of the evidence of its vulnerability to our missile system. There was no justification for
building this dog except maybe because it was being built in
LBJ's backyard, Fort Worth, Texas.

In fairness to McNamara, the Air Force was pushing for the
TFX. They wanted a tactical fighter-bomber that could be
used in big numbers in a ground war like Vietnam. The Blackbird was too revolutionary and too costly to fly regularly in
harm's way. The Air Force high command worried that it
would be shot down and its technological secrets fall into enemy hands. But General LeMay won a partial concession
from McNamara, and we received a contract for six two-man
reconnaissance versions of the Blackbird to be built exclusively for the Air Force. The plane would be ultimately designated as the SR-71. It was larger and heavier than the CIA one-
pilot model that carried only cameras. The Air Force model
would be packed with both cameras and supersophisticated
electronic eavesdropping equipment. The Skunk Works eventually would build thirty-one of them before this amazing airplane was finally, and some would say prematurely, retired in
the winter of 1990.

By then the Blackbird had become a legend as an incredibly
effective operational surveillance aircraft that could safely
overfly the most hostile and dangerous territory at will. But
the airplane and its operations were kept so secret that few
inside or outside our government knew it was flying. But the
Russians knew. So did the North Koreans, North Vietnamese,
and Chinese. And there was nothing they could do to stop it.

11

REMEMBERING HABU

MORE THAN THIRTY YEARS after its first flight, the
Blackbird's records will not soon be surpassed: New York to
London in one hour and fifty-five minutes; London to L.A. in
three hours and forty-seven minutes; L.A. to Washington in
sixty-four minutes. The Blackbird was 40 percent faster than
the Concorde, which first flew seven years later, and in 1964, its
creation won Kelly Johnson his second Collier Trophy, aviation's most prestigious award. He had won his first Collier five
years earlier, for building the world's first supersonic fighter,
the F-104. No one else in the industry had ever won two Collier
trophies, and that record will probably endure, too.

President Johnson awarded Kelly the Medal of Freedom,
our nation's highest civilian award, in 1967, not long after authorizing the first Blackbird overflights of North Vietnam in
May of that year. The president wanted hard evidence to back
up rumors that the North was receiving surface-to-surface
long-range ballistic missiles from the Russians that could
reach Saigon. Two CIA-piloted Blackbirds, flown by the
agency out of Kadena Air Base in Okinawa that summer, were
dispatched by presidential command to find out what was
really happening. They covered the whole of North Vietnam
photographically and found no evidence whatever of the presence of ground-to-ground missiles. Kelly joked that LBJ

bestowed the Medal of Freedom on him more in expression of
his relief than in gratitude.

For budgetary purposes LBJ ordered the CIA out of the spy
plane business in May 1968, and from then on all the missions
involving Blackbirds were conducted entirely by the Air Force
in its two-seater Blackbird, the SR-71. The second man, assist-
ing the pilot in the first cockpit, was the Reconnaissance Systems Officer, seated in the separate rear cockpit, who operated
all the avionics, as well as the nonautomatic cameras and radar frequency recording systems.

I won the American Institute of Aeronautics and Astronau-
tics Award in 1972 for designing the Blackbird's propulsion
system. But because the airplane operated in such secrecy for
the Air Force, and its training flights were conducted over
least-populated areas at tremendously high altitudes, few
Americans ever saw it fly and the public was only vaguely
aware of its existence. After LBJ officially announced the
Blackbird's creation in 1964, the Air Force was allowed to fly
the airplane for official speed and altitude records over closed
courses in 1965. The Blackbird established a new speed record
of 2,070 mph and an altitude record of 80,257 feet, even
though, over the twenty-five years it was operational, it rou-
tinely broke these records many times over while outclimbing
and outspeeding missile attacks. On one operational flight in
1976, a Blackbird actually reached 85,068 feet while flying at
2,092 mph.

Even now, many years after the fact, the public remains
oblivious to the harrowing and dangerous missions the airplane flew on an almost daily basis for more than a quarter
century. Many rumors still surround the "routine" penetration
overflights by the Blackbirds over such heavily defended denied territories as North Korea, North Vietnam, Cuba, Libya,

and the whole of the Soviet and Eastern bloc border, including intensive surveillance of the Russian nuclear submarine pens in the far frozen north. The Blackbird also performed a daily surveillance flight across the length of the demilitarized zone in Korea. General Larry Welch, as the blue-suiters' chief of staff, recalled having lunch with his South Korean counterpart near the DMZ when suddenly all the dishes rattled and the room rocked with a loud kaboom from the Blackbird's sonic boom. The Korean general smiled at Welch and sighed with satisfaction, "Ah, so."

We in the Skunk Works believed that the airplane's height and speed, as well as its pioneering stealthy composite materials applied to key areas of its wings and tail, would keep it and its crew safe, but we fortified that belief by adding a special fuel additive, which we nicknamed "panther piss," that ionized the furnace-like gas plumes streaming from the engine exhausts. The additive caused enemy infrared detectors to break up incoherently. We also implanted a black box electronic counter-measure in the airplane's tail called Oscar Sierra (the pilots called it "Oh, Shit," which is what most of them exclaimed when an ECM system activated at the start of an enemy missile attack). This ECM confused and distracted missile radar and kept it from locking on.

Time and again, the airplane and its counter-measure equipment proved their worth over North Vietnam, Cuba, and northern Russia. But mostly we could just outspeed any homing missile that would have to be led at least thirty miles ahead of its target to reach the Blackbird's altitude of sixteen miles high and at 2,000 mph–plus speeds. Most missiles exploded harmlessly two to five miles behind the streaking SR-71. Often the crew was not even aware they had been fired upon.

The Blackbirds flew 3,500 operational sorties over Vietnam and other hostile countries, had more than one hundred SAM SA-2 missiles fired at them over the years, and retired

gracefully in 1990 after twenty-four years of service as the only
military airplane never to be shot down or lose a single crew-
man to enemy fire. Which was truly amazing because the
Blackbird and its crews continuously drew the most dangerous missions. At such tremendous flying speeds, the margin for judgmental or mechanical error was zero, and at times
crews flew fourteen-hour round-trip flights more than halfway around the world, from their base in Okinawa, for example, to the Persian Gulf, providing fast, urgently needed
intelligence estimates on missile emplacements along the Iranian coast.

The complexity and duration of some of these missions defied belief — with ten or more air refuel rendezvous strung
out along the tens of thousands of miles of the typical long-
range Blackbird route. One screwup could result in tanks
going dry, a crashed airplane, and a lost crew. But it never
happened. Not once during a total of 65 million miles of flying, mostly at three times the speed of sound.

Like the U-2, the Blackbird was not an easy airplane to fly. It
demanded a pilot's total concentration and was unforgiving of
even smallish mistakes. The recollections of pilots and crews
attest to the awe and challenge of flying at speeds almost beyond human comprehension. Personnel were assigned to the
9th Strategic Reconnaissance Wing out of Beale Air Force
Base, near Sacramento, California, composed of thirteen two-
man crews flying nine active Blackbirds and supported closely
by a fleet of twenty-five KC-135s, the huge fuel tankers that
fanned out across the world to rendezvous with thirsty Blackbirds for air-to-air refueling. The crews who flew these missions were selected by SAC from the top of their list: of the
first ten pilots chosen, nine ultimately became generals. To a
man, all the pilots and their Reconnaissance Systems Officers
(RSOs) in the second cockpit regarded their tours as members
of the elite Blackbird unit as the highlight of their Air Force
careers.

Other Voices
Colonel Jim Wadkins
(Pilot)

I had 600 hours piloting Blackbird, and my last flight was just as big a thrill as my first. At 85,000 feet and Mach 3, it was almost a religious experience. My first flight out of Beale in '67, I took off late on a winter afternoon, heading east where it was already dark, and it was one of the most amazing and frightening moments going from daylight into a dark curtain of night that seemed to be hung across half of the continent. There was nothing in between — you streaked from bright day and flew into utter black, like being swallowed up into an abyss. My God, even now, I get goosebumps remembering. We flew to the east coast then turned around and headed back to California and saw the sun rising in the west as we reentered daylight. We were actually outspeeding the earth's rotation!

Nothing had prepared me to fly that fast. A typical training flight, we'd take off from Beale, then head east. I'd look out and see the Great Salt Lake — hell of a landmark. Then look back in the cockpit to be sure everything was okay. Then look out again and the Great Salt Lake had vanished. In its place, the Rockies. Then you scribbled on your flight plan and looked out again — this time at the Mississippi River. You were gobbling up huge hunks of geography by the minute. Hell, *you're flying three thousand feet a second!* We flew coast to coast and border to border in three hours fifty-nine minutes with two air-to-air refuelings. One day I heard another SR-71 pilot calling Albuquerque Center. I recognized his voice and knew he was flying lower than me but in the vicinity, so I called and said, "Tony, dump some fuel so I can see you." In only a couple of blinks of an eye, fuel streaked by underneath my airplane. He was like one hundred and fifty miles ahead of me.

One day our automatic navigation system failed. Ordinarily that's an automatic abort situation, but I decided to try to fly without the automatic navigation. I advised the FAA I was going to try this and to monitor us and let us know where we were if we got lost. I quickly learned that if we started a turn one second late, we were already off course, and if my bank angle wasn't exact, I was off by a long shot. I started a turn just below L.A. and wound up over Mexico! I realized right then that we couldn't navigate by the seat of our pants. Not at those incredible speeds.

I remember when a new pilot flying the SR-71 for the first time out of Beale began shouting "Mayday, Mayday" over Salt Lake City. "My nose is coming off!" My God, we all panicked and cranked out all the emergency vehicles. The guy aborted, staggered back to Beale. All that really happened was that the airplane's nose wrinkled from the heat. The skin always did that. The crew smoothed it out using a blowtorch. It was just like ironing a shirt.

My favorite route was to refuel over the Pacific right after takeoff, then come in over Northern California going supersonic, flying just north of Grand Forks, North Dakota, then turn to avoid Chicago, swing over Georgia, then coast out over the Atlantic, then refuel over Florida, west of Miami, then head straight back to Beale. Total elapsed time: three hours twenty-two minutes. Take off at nine or ten in the morning and land before two in the afternoon, in time to play tennis before cocktail hour.

As time went on we were being routed over least-populated areas because of growing complaints about sonic booms. One of them came straight from Nixon. One of our airplanes boomed him while he was reading on the patio of his estate at San Clemente. He got on the horn to the chief of staff and said, "Goddam it, you're disturbing people." One little community named Susanville, in California, sat right in a valley and was in the path of our return route to Beale. The sonic boom would echo off the hills and crack windows and plaster. We had the townspeople in, showed them the airplane, appealed to their patriotism, and told them the boom was "the sound of free-dom." They lapped it up.

Walt W. Rostow
(President Johnson's national security adviser from 1966 to 1968)

The Blackbird reconnaissance missions over North Vietnam, which began in late 1966, were invaluable to the president. We learned precisely the locations of missile and antiaircraft batteries, what ships were in the harbor unloading, and obtained up-to-date targeting intelligence for our bombing missions. Without these Blackbird overflights of Haiphong and Hanoi, President Johnson would never have allowed any tactical air operations in the North because he was extremely sensitive — I think in some ways, overly sensitive — to the possibilities of a bomb accidentally hitting a Chinese or Russian ship while it was unloading in the harbor, and he also was determined to keep civilian damage and casualties to a minimum. So he demanded frequent Blackbird missions — two or three every week — to supply him with the latest intelligence, since he usually chose the targets personally and insisted on approving each and every raid in the North. The military offered their priorities for targets, and the questions he raised about proposed targets were always the same: What were the military consequences of taking out that particular target? What were the expected losses in men and airplanes on this particular raid? And how much secondary civilian damage was anticipated? Before signing off on a mission he calculated in his own mind whether the anticipated losses were worth the anticipated gains. And the Blackbird photos were the decisive factors in helping him to make up his mind.

On January 23, 1968, the North Koreans caught us by surprise by boarding a naval surveillance ship called the *Pueblo* while it was in international waters. We were really caught short not knowing the fate of the crew or the ship. We figured they planned that incident to divert us at the time of the Tet Offensive in South Vietnam, but we would be damned if we'd let them get away with it. The whole country was up in arms over this incident. The president was considering using airpower to hit them hard and try to shake our crewmen loose. But when we cooled down, we had to suck in our gut and hold back until we were certain about the situation. Dick Helms, the director of the CIA, urged the president to authorize Blackbird flights to try to locate the missing ship. LBJ was reluctant to overfly North Korea and offer the tempting target of the Blackbird, possibly provoking an even greater international incident. But he was assured that Blackbird could photograph the whole of North Korea, from the DMZ to the Yalu River, in less than ten minutes, and probably do so unobserved by air defense radar. Which is precisely what happened. The Blackbird quickly located the captured *Pueblo* at anchor in Wonson harbor only twenty-four hours after it was boarded by the North Koreans. So we had to abandon any plans to hit them with airpower. All that would accomplish would be to kill a lot of people, including our own. But the Blackbird's photo take provided proof that our ship and our men were being held. The Koreans couldn't lie about that, and we immediately began negotiations to get them back.

Captain Norbert Budzinske
(Air Force RSO)

We trained a year, acquired a hundred hours of flying time, before we flew out to giant Kadena Air Base at Okinawa in early 1968. The trip took about six hours by SR-71, fourteen hours if you flew over in a tanker. This was for real. The slower and lower U-2s couldn't make it flying over North Vietnam with all those SAM sites. That mission became ours. We flew five times a week on average. Also, missions over North Korea and to Soviet naval facilities, north of Japan. We operated

three airplanes and lived isolated from the main base population in a remote corner of Kadena that had been set up for the
original CIA pilots and crew. But very quickly the islanders
learned we were there and would go up on the hill overlooking
the base to catch a glimpse of our bird, which shook the whole
island taking off. The islanders nicknamed our airplane Habu,
after a black, extremely deadly pit viper. Habu, like the Blackbird, was indigenous to Okinawa and the name stuck. We carried that name on our shoulder patches. To this day, that hill is
still called Habu Hill by the islanders.

The minute we showed up at Kadena, so did two Soviet
trawlers, which fed our takeoff times and position to their
friends in North Vietnam and put them on the alert. But we
were cocky. We timed our flights not for surprise, but to
achieve the most favorable sun angle for our pictures, usually
taking off between ten in the morning and noon. We made a
big turn and banked to head for Hanoi, exposing ourselves on
the side to those giant Tall King long-range radars that the Chinese installed on Hainan Island. They knew we had two primary targets — Haiphong and Hanoi — so it was just a matter
of timing us and waiting for us to fly over their missile sites. We
collected so much photo and electronic intelligence data from
each mission that our intelligence people were swamped by it
and simply couldn't keep up. We photographed 100,000 square
miles of terrain per hour, from which selected target images
could be blown up more than twenty times. Some of the cameras were operated by me as the RSO in the backseat and
others were automatic. The cameras were very fast and pre-
programmed. We could provide both horizon-to-horizon coverage and close-up telescopic work that would let you see
down a flea's throat.

Okinawa is one hell of a long way from Vietnam. Our missions averaged more than six thousand miles and took more
than four hours. The logistics involved were staggering because we had refueling tankers strung out from Kadena to

Thailand. So we usually flew two airplanes, using one as a spare in case the mission plane had to crap out. Having all those tankers in the sky was expensive. Typically, we'd hit a tanker fifteen minutes after takeoff, head on down between Taiwan and the Philippines, to Cam Ranh Bay, then turn north over North Vietnam. We'd overfly the north in eight to twelve minutes, take in both Haiphong and Hanoi, staying out of range of the SA-2 missile batteries down below. Then we'd do a double-looper, going over the north again, before refueling over Thailand. We overflew both north and south in Vietnam, did Laos and Cambodia, flew over both north and south in Korea.

Sometimes after a mission, I'd get a look at our photo take to see how well we did. It was unbelievable! You could actually see down the open hatches of a freighter unloading in Haiphong's harbor. In fact, the photo interpreters claimed that they could tell what was down in those hatches, it was that sharp and clear from 85,000 feet. They'd blow up our photos to the size of a table.

Frankly, I had the feeling that there were things the hostiles made no effort to hide. For example, North Korea. Took us about ten minutes to overfly then hit a tanker and do it again at a different latitude. These missions had to be cleared at the highest level. We concentrated on missile sites, port facilities, any unusual movement. Those folks never moved their missile sites. They wanted us to see them, record them. Not cause tensions or apprehensions by hiding them or moving them around. Same with the Chinese. We overflew them two or three years, but only with direct orders from the commander in chief. Those guys wanted us to know they had the bomb. That was clear.

We knew we'd probably get shot at, but it wasn't a big worry because at our height an SA-2 missile simply didn't have the aerodynamic capability to maneuver once we started twisting and turning to get away from it. Still, when a little warning light came on in my cockpit to report that the boys downstairs had launched one up at me, I tightened up for sure. I personally never saw a missile coming up, but others did. One of our pilots, Bill Campbell, saw three missiles explode at his altitude, but a couple of miles behind. He actually watched them being launched and caught it on film. He said those damn missiles looked as big as telephone poles as they lifted off.

Later on, in the late 1970s, when the Russians developed their powerful SA-5 ground-to-air missile that could have knocked us down, they never tried to use it against the Blackbird. That missile was so enormous it looked like a medium-range intercontinental ballistic missile sitting on its pad. Made me queasy just looking down at it through my telescopic sight. But my theory was that the hostiles realized that reconnaissance flights were actually stabilizing. We knew what they were looking at and they knew what we were looking at. If they denied us, we'd deny them. And then everyone would get the jitters. In this game, you didn't deny access unless you were ready to get serious about preventing it.

Lt. Colonel Buz Carpenter
(Air Force pilot)

We were flying over North Vietnam at 82,000 feet when suddenly I had to do a quick maneuver to keep us from colliding with a weather balloon that whisked past us just off my wingtip. At that very moment, we experienced an unstart in the left engine, which knocked me and my RSO all over our respective cockpits. My RSO looks out and sees the left engine trailing fuel, and he thinks we've been hit by a SAM. "Buz, we've been hit," he shouts into his open mike. I always suspected that we were more closely monitored on these flights than we probably realized by special RC-135 snooper airplanes packed with powerful electronics. And now, I'm certain, because by the time we made our second pass over the North, the White House Situation Room had already been wrongly notified that my SR-71 had been hit over Hanoi.

Major Butch Sheffield
(Air Force RSO)

Just before I deployed to Kadena in the fall of '69, we were
tasked to fly up to the Arctic Circle and check on some sus-
picious activity on a tiny Russian island. I was then stationed
at Beale, living at home with my wife. We took off after breakfast, refueled over Alaska, headed north for fifteen hundred
miles. It was scary being over the most forbidding area of the
world. If you had to eject, you were finished. Anyway, we
reached this island, turned on the recorders and the cameras,
then turned around, hit another tanker, and flew back into
Beale. I was home in time for dinner and my wife never knew
where I'd been that day. She assumed it was just another day at
the office. At dinner, I almost burst out laughing thinking of
her reaction if she had known I'd spent the day flying up and
back to the Arctic Circle!

I had another very odd mission, like that one, in the early
1970s, while stationed at Kadena. We were tasked to fly
against the very formidable and new Soviet SA-5 missile site
that had been constructed at Vladivostok, their big naval base
in the Sea of Japan, at a time when the Russians were con-
ducting a huge naval exercise right off the coast. We were to
fly to this dangerous site late Sunday night, hoping that we
would find their most junior officers on duty, who would snap
at the bait and turn on their radar and we could measure
the frequency, the pulse repetition intervals, and a lot of
other vital technical details that could be used to develop
counter-electronics against this monster. A more experienced
officer might figure out what we were really up to and stay
dark.

The National Security Agency put aboard a special recording package for this particular mission, and we flew right
down the throat of that site, so that it seemed certain that we
were going to overfly Soviet territory. As we came in, radars

from dozens of Soviet naval ships on that training exercise switched on. And at the last second, we pulled a sharp turn and avoided any overflight of Soviet airspace. But the take was awesome. In all, we got nearly three hundred different radar transmissions recorded, including the first SA-5 signals obtained by our side. Meanwhile, I had plenty of problems to cope with before we landed safely.

After turning, we headed toward Japan. During that big turn, the oil pressure on the left engine began falling and rapidly dropped to zero. We stayed on a southern heading but shut down that engine and flew on one engine against awful headwinds at fifteen thousand feet as we approached the North Korean coastline. We were just struggling to maintain altitude and didn't realize it until later that the North Koreans had scrambled fighters against us, and that the South Koreans had scrambled their fighters to get between us and the North Koreans and defend us. We had no choice but to put in to a South Korean air base at Taegu. I called the field, but they refused permission to land. Field closed, they said. I said, "I've got to land. Turn on your lights." We came in and just sat there, surrounded by dozens of people in black pajamas with machine guns. My pilot said to me, "Butch, you sure we landed in *South Korea?*"

On November 22, 1969, I flew my first North Korea mission. I was uptight because North Korea was very heavily defended, and on this particular mission we went to every known SAM site in North Korea, all twenty-one of them, and we crisscrossed them, made a big 180-degree turn that took us right across the Chinese border, then came back right down the center of North Korea, then made another 180-degree turn across the demilitarized zone. We crossed North Korea eight to ten times on that mission, covered the entire country. As we finished up and were turning to go home, a right-generator fail light came on. I tried to reset it, but it was no go, so we ended up making an emergency landing in the south and caused a big stir and fuss.

In '71, we were tasked to fly three Blackbirds over North Vietnam, which was highly unusual. All other missions used only one airplane at a time. We took off first, refueled over Thailand, and headed north, with the other two planes following. The plan was for us to crisscross over Hanoi in thirty-second intervals at 78, 76, and 74 thousand feet respectively, at a certain point, which we later learned was over the Hanoi Hilton, the infamous POW prison, and deliver sonic booms, one after another. Later on, the vice commander of our squadron hinted that the purpose of the mission was to send a signal to the POWs, a fact I've never been able to confirm. Many years later, I talked to POWs who were in that prison at the time and they heard the sonic booms thirty seconds apart but insisted that they didn't know what in hell it meant.

Fred Carmody

I ran the SR-71 operation for Lockheed at Mildenhall, an RAF base in Suffolk, from 1982 to 1989. We had two SR-71s, and when I went in with eighteen Skunk Works mechanics, I replaced eighty Air Force mechanics who had suffered eight aborted missions in a row. We took over and immediately logged eight successful flights in a row and had those blue-suiters scratching their heads.

The Mildenhall operation was revealed by the prime minister, Mrs. Thatcher, about two years after we got there. Our main deployment was to fly twice a week up to the northern extremes of the Soviet Union to the big naval base of Murmansk, on the Kola peninsula, to keep a sharp eye on their nuclear sub pens. That place was so remote it took us three air-to-air refuelings to get there. Which is why they put their subs there to begin with. The place was very heavily defended by fighters and missiles, but only the Blackbird could fly it safely. They always detected us, but they couldn't stop us. We flew right up against them but didn't actually penetrate Soviet airspace. That was a big no-no, by presidential orders. In cases of emergency, engine problems and such, we had to put in to Norway — that happened three times. We took pictures of the subs in their pens, saw which ones were occupied, which empty. Those nuclear subs were a potential threat to our mainland because they carried Polaris-type nuclear missiles that could be fired offshore and hit Washington and New York. So we kept a close count and tracked the subs from the moment they left their pens. The pictures were so clear we could even tell the size of the sub's screws under the water, count the missile silos on the decks. We took both radar pictures and regular pictures, tested radar frequencies by penetrating their air defense systems, and then did the same thing on long flights down the Baltic coast of East Germany and Poland.

The Soviets would scramble their MiG-25s against us, but they got up to only sixty thousand feet, then fell off. One of them got up to Mach 3 on a zoom, then fell off. Those MiGs could reach max speeds of Mach 2.5 to 2.7, which wasn't nearly good enough to catch the SR-71. They also fired missiles at us from time to time, but we used our jammers very effectively. The purpose of the Baltic coastal flights was to gather radar and electronic intelligence as well as obtain good photos of military facilities along the border of the Eastern bloc. On a typical mission, we could get invaluable intelligence on as many as fifty different radar and missile tracking systems deployed against us in the Baltic and Barents Sea regions.

Our guys out of England regularly overflew the Mediterranean, from Greece to Tunisia. We were far superior to any damned spy satellite. You wanted a picture at 10:01 a.m. Sunday, anywhere in the world, we'd go out and get it for you. We covered the world with a handful of airplanes.

Lt. Colonel D. Curt Osterheld
(Air Force RSO)

My first mission to the Soviet sub pens in the Baltic was in December 1985. At eighty-five thousand feet, the winter view of the northern Russian port areas was magnificent. The land was absolutely white in every direction, and the sea was frozen solid, too, except for the dark black lines where icebreakers had been used and I could see subs moving along the surface, heading for deeper waters. In the polar cold, we pulled engine contrails, which made us a clear target in the cloudless sky. Way down below I could see the spiral contrails of Soviet fighters, scrambled because of us. At that time of year the sun was very low on the horizon even at noon. I had lowered the curtain to cover the left window so I could watch the radar and defensive systems panels in the backseat without having the sun directly in my eyes. As we approached the sub pens we received a "condition one" alert over the high-frequency radio that got our immediate attention. An instant later, the defensive systems panel began lighting up, indicating we were being tracked and engaged by SAM missiles and there might be a missile on the way. I advised my pilot to accelerate out to maximum Mach. I thought, If they really fired a missile, it should be here by now. At that instant, the Velcro that held the sun screen curtain in place let go, and my cockpit was lit with brilliant light. For a split second, I thought, They got us! I let go with a shriek of terror. So much for the image of the fearless flyer. Scared crapless by a blinding flash of winter sunlight.

Lt. Colonel William Burk Jr.
(Air Force pilot)

In the fall of '82, I flew from Mildenhall on a mission over Lebanon in response to the Marine barracks bombing. President Reagan ordered photo coverage of all the terrorist bases in the region. The French refused to allow us to overfly, so our

mission profile was to refuel off the south coast of England, a
Mach 3 cruise leg down the coast of Portugal and Spain, left
turn through the Straits of Gibraltar, refuel in the western
Mediterranean, pull a supersonic leg along the coast of Greece
and Turkey, right turn into Lebanon and fly right down main
street Beirut, exit along the southern Mediterranean with another refueling over Malta, supersonic back out the straits, and
return to England.

Because Syria had a Soviet SA-5 missile system just west of
Damascus that we would be penetrating (we were unsure of
Syria's intentions in this conflict), we programmed to fly above
eighty thousand feet and at Mach 3 plus to be on the safe side,
knowing that this advanced missile had the range and speed to
nail us. And as we entered Lebanon's airspace my Recon Systems Officer in the rear cockpit informed me that our defensive
systems display showed we were being tracked by that SA-5.
About fifteen seconds later we got a warning of active guidance
signals from the SA-5 site. We couldn't tell whether there was
an actual launch or the missile was still on the rails, but they
were actively tracking us. We didn't waste any time wondering,
but climbed and pushed that throttle, and said a couple of
"Hail Kellys."

We completed our pass over Beirut and turned toward
Malta, when I got a warning low-oil-pressure light on my right
engine. Even though the engine was running fine I slowed
down and lowered our altitude and made a direct line for England. We decided to cross France without clearance instead of
going the roundabout way. We made it almost across, when I
looked out the left window and saw a French Mirage III sitting
ten feet off my left wing. He came up on our frequency and
asked us for our Diplomatic Clearance Number. I had no idea
what he was talking about, so I told him to stand by. I asked my
backseater, who said, "Don't worry about it. I just gave it to
him." What he had given him was "the bird" with his middle

finger. I lit the afterburners and left that Mirage standing still. Two minutes later, we were crossing the Channel.

Major Randy Shelhorse
(Air Force RSO)

On October 28, 1987, I flew from Okinawa to Iran and back — an eleven-hour mission into the Persian Gulf. The purpose was deadly serious. The Reagan White House wanted to know whether or not the Iranians were in possession of Chinese Silkworm missiles that could be fired against shipping in the Straits of Hormuz. These missiles posed a threat against oil tankers, and the Blackbird overflights clearly showed that the Iranians indeed had Silkworms in place along the coast. Knowing their exact location, we were able to forewarn the Navy and also deliver some very direct private warnings to the Iranians about what they could expect if they fired one of those Silkworms at shipping. Our government was escorting tankers in and out through the narrow straits to ensure their safety.

Our flight was code-named Giant Express. And that it certainly was. We flew halfway around the world. After leaving Okinawa, we headed out toward Southeast Asia, and then, south of India, for our second refueling. Five hours after taking off, we approached the Gulf region and could actually see the Navy escorting a line of tankers down below. While in acceleration after the third air refueling, I received the following radio call: "Unidentified aircraft at forty thousand feet, identify yourself or prepare to be engaged." I immediately transmitted on another UHF frequency specifically designated for this sortie that I was "an Air Force special." I received no further response. A few months later, an Iranian civilian airliner was shot down by the U.S. Navy in the Gulf. While reading the press accounts of the incident, I recognized the identical transmission to the Airbus that I had received, probably from that very same missile frigate down below. I sometimes wonder

what would have happened had I chosen not to respond to
their identification demand and maintained my radio silence.

Some of the riskiest missions had to be personally cleared by
the president and were undertaken at moments of high drama
and international tensions when the chief executive's need to
know what was happening inside denied or hostile territory
was so explicit that issues of war and peace hung in the
balance.

For example, during the early hours of the 1973 Yom Kippur War, when the Arab armies caught the Israelis by surprise
and scored quick victories on three separate fronts, President
Nixon was informed that the Russians had repositioned their
Cosmos satellite to provide their Arab clients with real-time
overflight intelligence showing troop positions and
deployments — a huge tactical advantage. Nixon ordered
Blackbird overflights to provide these same kinds of real-time
war zone overviews to the Israelis and level the battlefield.
However, the British government, afraid of offending the
Arabs, refused to allow the Blackbird mission to leave or land
at their Mildenhall base, so we flew nonstop to the Middle
East from a base in upstate New York — a twelve-thousand-
mile round-trip in less than half a day. By the following day,
Blackbird's photo take was on the desk of the Israeli general
staff.

During one of the most tense moments of his presidency,
Reagan ordered the Air Force Blackbirds to mount deep-
penetration flights along the Polish-Soviet border, in January
1982, following the Polish government's brutal crackdown
against the Solidarity reformers. Poland's Communist ruler,
General Jaruzelski, cut communications with the West and
declared martial law; the White House was deeply concerned
that the Kremlin not only had ordered these drastic moves,
but was about to commit troops to crush the uprising as they
had done in Czechoslovakia in 1968. Much to Reagan's relief,

the overflights revealed no Soviet troop movements or any evidence of a military buildup along the border.

The Blackbird flew again by direct presidential order during Operation Eldorado Canyon, the April 1986 air raid against Muammar Kaddafi of Libya, in direct retaliation for a terrorist attack against a Berlin nightclub frequented by U.S. servicemen. The Blackbirds provided post-raid reconnaissance and damage assessments. These overflights, only six hours following the raid, were extremely dangerous since Libya's entire air defense system was on maximum alert and eager to bring down the prestigious Blackbird as a prized trophy. Dozens of missiles were launched, but none came close. The photo take was transmitted to the Situation Room an hour after safe landing back in Britain, completing a six-hour, twelve-thousand-mile round-trip, and a few hours later, table-size blowups of bomb damage were shown to Secretary of Defense Weinberger. One photo revealed that an errant bomb had accidentally hit an underground ammunition storage facility not far from the presidential palace. The Libyans thought that the hit was intentional and were stunned at our apparent intimate knowledge of their secret storage facilities. "How did the Americans know?"

In my view, shared by many blue-suiters, this marvelous airplane should still be operational but, alas, that was not to be. One of the most depressing moments in the history of the Skunk Works occurred on February 5, 1970, when we received a telegram from the Pentagon ordering us to destroy all the tooling for the Blackbird. All the molds, jigs, and forty thousand detail tools were cut up for scrap and sold off at seven cents a pound. Not only didn't the government want to pay storage costs on the tooling, but it wanted to ensure that the Blackbird never would be built again. I thought at the time that this cost-cutting decision would be deeply regretted over the years by those responsible for the national security. That decision stopped production on the whole series of Mach 3 aircraft for the remainder of this century. It was just plain dumb.

But the Air Force decided that the twenty or so SR-71s remaining in service from the original procurement of thirty-one aircraft could suffice through the end of this century. In fact, a study by the Defense Science Board review, in 1984, concluded that the Blackbird's outer titanium skin, annealed by heat on every flight, was actually stronger than when first delivered more than a decade earlier and would last another thirty years. The blue-suiters decided to invest $300 million in updating the airplane's electronics with digital flight controls and a new weather-penetrating synthetic-aperture radar system, as well as refurbishing its power systems. But General Larry Welch, the Air Force chief of staff, staged a one-man campaign on Capitol Hill to kill the program entirely. General Welch thought sophisticated spy satellites made the SR-71 a disposable luxury. Welch had headed the Strategic Air Command and was partial to its priorities. He wanted to use SR-71 refurbishment funding for development of the B-2 bomber. He was quoted by columnist Rowland Evans as saying, "The Blackbird can't fire a gun and doesn't carry a bomb, and I don't want it." Then the general went on the Hill and claimed to certain powerful committee chairmen that he could operate a wing of fifteen to twenty F-15 fighter-bombers with what it cost him to fly a single SR-71. That claim was bogus. So were claims by SAC generals that the SR-71 cost $400 million annually to run. The actual cost was about $260 million.

SR-71 operations were not cheap; they could not fly at cruise speed longer than an hour and a half without requiring the costly and complex planning of air-to-air tanker refueling. And what really annoyed the blue-suiters was the fact that while the Navy, the State Department, and the CIA shared in

the intelligence takes acquired on SR-71 missions, none of
these users helped to defray the operational expenses. In 1990,
Defense Secretary Dick Cheney decided to retire the airplane
and end the program. Some of our friends in Congress, like
Senator John Glenn, were bitter about the decision, after a
two-year struggle to keep it from happening. Senator Glenn
warned, "The termination of the SR-71 is a grave mistake and
can place our nation at a serious disadvantage in the event of
a future crisis." We had more than forty members of Congress
actively seeking to keep the program alive, headed by Senator
Sam Nunn, the powerful chairman of the Senate Committee
on Armed Services. But Cheney prevailed. I was a fan of his,
thought he was an outstanding DOD Secretary, but I agreed
with Admiral Bobby Inman, then the former director of the
National Security Agency, who commiserated with me over
Cheney's decision. "Satellites will never fully compensate for
the loss of the Blackbird," Bobby told me. "They have nothing
in the wings to replace it and we may be in for some nasty
surprises and a whole new set of intelligence problems because of this." A few days after Iraq invaded Kuwait in 1990, I
called General Michael Loh, Air Force vice chief of staff, and
told him that I could have three Blackbirds ready and operational in ninety days to overfly the region. I also could supply
qualified pilots. The last three Blackbirds were being used by
NASA for high-altitude flight tests. My idea was to provide the
blue-suiters with a total package — airplanes, pilots, and
ground crews — for a cost of about $100 million. The air-
planes would be indispensible providing surveillance over
Iraq, and I had another idea, too. "General," I said, "We could
fly over the rooftops of Baghdad at Mach 3 plus at prayer time
and sonic-boom the bastards. Just think how demoralizing
that would be for Saddam." General Loh said he would get
back to me. About a week later I received a call saying that
Dick Cheney had vetoed the idea. The secretary felt that there

was no such thing as a one-time-only role for the Blackbird. "Once we let this damned airplane back in, we'll never get it back out," he told General Loh.

Other Voices
Ed Yeilding
(Air Force pilot)

When Congress approved the decision to retire the SR-71, the Smithsonian Institution requested that a Blackbird be delivered for eventual display in the Air and Space Museum in Washington and that we set a new transcontinental speed record delivering it from California to Dulles. I had the honor of piloting that final flight on March 6, 1990, for its final 2,300-mile flight between L.A. and D.C. I took off with my backseat navigator, Lt. Col. Joe Vida, at 4:30 in the morning from Palmdale, just outside L.A., and despite the early hour, a huge crowd cheered us off. We hit a tanker over the Pacific then turned and dashed east, accelerating to 2.6 Mach and about sixty thousand feet. Below stretched hundreds of miles of California coastline in the early morning light. In the east and above, the hint of a red sunrise and the bright twinkling lights from Venus, Mars, and Saturn. A moment later we were directly over central California, with the Blackbird's continual sonic boom serving as an early wake-up call to the millions sleeping below on this special day. I pushed out to Mach 3.3.

From Kansas City eastward we were high above a cirrus cloud undercast but savored this view from above 97 percent of the earth's atmosphere — enjoyed witnessing for one final time the curvature of the earth, the bright blue glow just above the horizon, and the pitch-dark daytime sky directly overhead. High above the jet stream, the winds blew at only five mph, and we cruised smooth as silk.

We averaged 2,190 mph from St. Louis to Cincinnati, covering the distance in eight minutes, thirty-two seconds, a new city-to-city aviation record.

When we were abeam Washington at eighty-four thousand feet, I terminated the supersonic afterburner and began our descent. We had set two records: L.A. to D.C. in only sixty-four minutes and Kansas City to D.C. in twenty-six minutes. And we had set a new transcontinental speed record, covering 2,404 statute miles in only sixty-seven minutes, fifty-four seconds. It was also the first time that a sonic boom had traversed the entire length of our great country. Through the haze I saw the Dulles tower and flew above the waiting crowd at eight hundred feet, resisting the temptation to really go down on the deck for fear of blowing out the Dulles terminal windows with our powerful engine vibrations. I felt both tremendous elation and tremendous sadness. When we landed and climbed out, Ben Rich, head of the Skunk Works, was waiting below to shake my hand. I had met him once before, when I worked with him to plan a fly-by of this airplane directly over the Skunk Works on December 20, 1989, celebrating the twenty-fifth anniversary of the Blackbird.

I made three low passes over the complex of hangars and buildings that comprise the famous Skunk Works operation at Burbank, and Ben had trotted out every single worker to cheer and experience the thrill of seeing this incredible machine they had built sweeping in low over their heads. There were several thousand workers down below waving at us. On the last pass I performed a short, steep afterburner climb and rocked my wings in a salute. I heard later that men had cried.

12

THE CHINA SYNDROME

THE MOST SENSITIVE project during my years at the Skunk Works was begun in 1962 and code-named Tagboard. Inside the Skunk Works and out, fewer than one hundred people were involved or knew about it, even though the program began under President Kennedy and became operational under Richard Nixon. To this day, Tagboard remains mist-shrouded, in part because it was basically a failure. But still, the operations were truly spectacular: flying spy drones over Communist China's most remote and secret nuclear test facilities.

I was fascinated by drones. To me, a remote-controlled or preprogrammed pilotless vehicle was the pragmatic solution to spying over extreme hostile territory without worrying about loss of life or political embarrassments of the Francis Gary Powers variety. If the drone traveled high enough and fast enough, the enemy could not stop it — indeed might not even spot it. Several of us in the analytical section were drone boosters and from time to time tried to lobby Kelly Johnson into joining our fan club. But Kelly resisted. Drones, he argued, were too big and complex to be economically feasible or operationally successful. But over the years he gradually changed his mind. He was aware of the ominous crash development of nuclear and rocket weapons on the Chinese mainland and the loss of four Taiwanese U-2s, shot down while trying to film those sensitive sites. The most significant of these development facilities was Lop Nor, two thousand miles inland, practically to the Chinese border with Mongolia. Lop Nor was situated inside a two-thousand-foot depression twenty miles wide on an otherwise four-thousand-foot plateau. This rugged, isolated test facility was the primary target of U.S. intelligence, especially in the face of an aggressive and increasingly hostile Chinese foreign policy. The Chinese had split with Moscow, triggered border clashes with India and Tibet, and were causing major concerns in nearly every foreign capital around the world.

Lop Nor was two thousand miles away and a very tough round-trip for even the most experienced U-2 pilot. Those of us promoting the drone idea inside the Skunk Works argued with Kelly that it was the best way to overfly forbidding places like Lop Nor. We envisioned a delivery system where a drone would be piggybacked on top of the Blackbird and launched off the China coast, rocket up to 100,000 feet, and zip over Lop Nor at speeds faster than Mach 3, take its pictures, then turn around and fly back to the launch point, where on electronic command it would drop its film package by parachute to a waiting naval frigate in the sea below. The drone would then self-destruct. The technology involved was not only feasible but within our grasp.

Kelly got a negative reception to the drone idea from John Parangosky, who had replaced Dick Bissell at the CIA. The Air Force was only slightly more receptive. Nevertheless, Kelly found an ally in Brig. Gen. Leo Geary, director of special projects in the Air Force, who coordinated programs between the CIA and the blue-suiters. Geary obtained half a million dollars in seed money from "black project" contingency funds, and we put together a small team to plot and plan a design. I was the propulsion man. "This is a *most* peculiar situation," Kelly told us. "The agency has turned its back, so Lockheed might wind up launching this damned thing ourselves. I have no instructions from anyone in Washington, but I think I know what they want: we will try to get six-inch ground resolution photo-graphically, a range of at least three thousand nautical miles, a camera payload of 425 pounds, and a guidance system of about 400 pounds. We should detach the payload bay holding the camera, film, and guidance system and float it down by parachute. Make it reusable and save a bundle of money that way. That's it. Go to it."

The drone we designed had the flat triangular shape of a manta ray, was forty feet long, weighed about seventeen thousand pounds, would be built from titanium, powered by the same kind of Marquardt ramjet we once used for an experimental ground-to-air missile developed in the 1950s, called Bomarc. The drone had the lowest radar cross section of anything we had ever designed and could cruise faster than three times the speed of sound. It was equipped with a star-tracker inertial guidance system that could be constantly updated via computer feeds from the system aboard the mothership until the moment of launch. The system was fully automated, and the drone's steering was directed by stored signals to its hydraulic servo actuators. It was capable of a sophisticated flight plan, making numerous turns and twists to get where it was going, then repeating them in reverse to return to where it came from. The payload was detached on radio command after the mission and parachuted to a waiting cargo plane equipped with a Y-shaped catching device. After the nose detached, the drone self-exploded.

Kelly took our design to Washington in February 1963. His reception at the CIA was still unenthusiastic, mostly because the agency was already overextended on a huge secret budget for hardware, involving Blackbirds and spy satellites and feeling heat from congressional oversight committees. In fact, the Bureau of the Budget was rumored to be taking a hard look at the CIA's air wing, with an eye toward eliminating it entirely, which finally occurred by presidential command in 1966.

But Air Force Secretary Harold Brown was definitely interested in the drone concept, as much as for a way of delivering a nuclear bomb as for spying: they could deploy a nuclear weapon by drone three thousand miles from a hotly defended target and be impossible to stop. Apparently, the CIA got serious about our drone after learning about the Air Force interest. With General Geary acting as our champion, the agency decided to climb on board the drone project. On March 20, 1963, we were awarded a letter contract from the CIA, which would share funding and operational responsibilities with the Air Force. Ultimately, we built fifty drones for only $31 million before the project ended in 1971.

Tagboard now became the most classified project at the Skunk Works, even more secret than the Blackbird airplanes being assembled. So Kelly decided to wall off a section of the huge assembly building housing Blackbird, which already was as guarded as Fort Knox, to accommodate the new drone project. To get inside that walled-off section required special access passes and the shop workers immediately dubbed it Berlin Wall West. Unfortunately, I found myself spending more time inside that walled section than I had ever anticipated. But the technical problems were formidable, especially the attempt to launch a piggybacked drone from a mothership launch platform flying at three times the speed of sound. The drone would be sitting toward the top rear of the fuselage on a pylon. Expecting a drone to launch through the mothership's Mach 3 shock wave presented a monumental engineering challenge. And Kelly insisted that we launch at full power.

It took us nearly six months to work out some of the shock wave and engine problems with models in the wind tunnel, while other problems concerned with perfecting the guidance system, the cameras, the self-destruction system, the parachute deployment system, all loomed before us like monsters let out of some evil sorcerer's dungeon. But to Kelly the biggest sweat was guaranteeing a safe launch. I had never seen him so spooked. "Goddam it, I don't want to lose a pilot and an airplane testing this system. This will be the most dangerous maneuver in any airplane that I've ever worked on. And I don't want that damned drone flying out of control and crashing into the middle of downtown Los Angeles or Portland."

He kept postponing test launches and finally aimed for the first one on his birthday — February 27, 1965. But that first flight did not occur until we had solved dozens of complicated problems, thirteen months later. Bill Park finally took off in a Blackbird from our secret base with a drone sitting on top of his fuselage. Out over the California coast at 80,000 feet and at Mach 3.2, Bill ignited the drone, which launched perfectly and flew 120 miles out to sea before running out of fuel and crashing. One month later, a second launch was spectacularly successful. The drone flew 1,900 nautical miles at Mach 3.3, holding to its course all the way, and finally fell out of the sky when a hydraulic pump burned out.

On June 16, 1966, we attempted the third test launch of the drone piggybacking on an SR-71 Blackbird, a two-seater. Bill Park was our pilot, and in the second cockpit was Ray Torick, the launch operator. The Blackbird took off and headed for the California coast, just north of L.A., to launch over the naval tracking station at Point Mugu. The flight was a dandy. The drone flew 1,600 nautical miles, making eight programmed turns while taking pictures of the Channel Islands, San Clemente, and Santa Catalina from 92,000 feet at 4,000-plus mph. It did everything but eject the film package, due to electronic failure, which was a very fixable problem. So a few weeks later, on July 30, 1966, we repeated the same test flight over Point Mugu, just up the coast from Malibu. This time we launched at 3.25 Mach and — *catastrophe.*

The drone crashed into the fuselage of the Blackbird, which spun wildly out of control. Park and Torick both ejected with their pressure suits inflated. Bill Park was picked up in a life raft 150 miles at sea. Torick splashed down nearby, but rashly opened the visor of his helmet while he was paddling in the ocean, so that water flooded into his pressure suit through the neck ring and he sank like a stone. Our flight director, Keith Beswick, who was flying chase, had to go to a local mortuary and cut him out of the pressure suit so that the body could be properly prepared for burial.

We were shaken, but no one more so than Kelly Johnson, who was so upset at Torick's death that he impulsively and emotionally decided to cancel the entire program and give back the development funding to the Air Force and the agency. Several of us urged him to reconsider. The drone project was actually doing well. But Kelly seemed adamant: "I will not risk any more test pilots or Blackbirds. I don't have either to spare." Kelly could now turn to the SAC B-52 bomber as the vehicle to carry the drone. The B-52 was subsonic, which would decrease the dangers of an air launch. And it could launch our drone from about forty-five thousand feet, which would not cut very deeply into our bird's extended range.

Kelly flew to Washington and, shortly after Christmas 1966, met for over an hour with LBJ's deputy defense secretary, Cyrus Vance, who was enthusiastic about Tagboard and authorized the use of B-52s as the mothership. Vance told Kelly, "We need this project to work because our government will never again allow a Francis Gary Powers situation to develop. All our overflights over denied territory will either be with satellites or drones."

The Air Force supplied two specially equipped H Model B-52s, an eight-engine mothership that would carry a drone under each wing on a sixteen-and-a-half-hour, seven-thousand-nautical-mile flight, from Beale Air Force Base in central California to a launch point west of the Philippines, over the South China Sea. The drone's mission was to overfly China. It would launch from forty thousand feet and fire its solid-propellant booster rocket for eighty-seven seconds, soaring it to eighty thousand feet at Mach 3.3 plus, before falling away. The drone's guidance system would direct its three-thousand-mile journey. On the return leg of the mission, the drone was programmed to descend to sixty thousand feet and slow to Mach 1.6 before automatically ejecting a hatch containing its film, camera, and expensive telemetry and flight control components, which could be used again after successful parachute recovery.

By the winter of 1968 the Skunk Works and the Air Force teamed up to launch long-range test flights using the mighty B-52, which carried two drones, one under each wing, for a total carrying payload of twenty-four tons. The tests were conducted in Hawaii, with the drones launched thousands of miles across the open Pacific, overflying such places as Christmas Island or Midway, taking pictures of pre-programmed targets on these islands, then looping back toward Hawaii and culminating three thousand miles of precise flying with a rendezvous point above a recovery team of picket ships and cargo planes. Our track record on these so-called Captain Hook test flights was five successes and two failures over fourteen months. And by the fall of 1969 a special committee of CIA and Air Force analysts began recommending hot missions to the EXCOM (Executive Committee of the National Security Council). EXCOM approved and forwarded its recommendation to President Nixon. We received a go for a first mission over Lop Nor nuclear test range on November 9, 1969. A lone B-52 took off from Beale Air Force Base in the predawn hours carrying two Tagboard drones, one under each wing — in case one failed to ignite. After a single air-to-air refueling some twelve hours later, it reached its launch point at the fourteen-hour mark, and Tagboard successfully fired beyond range of Chinese early-warning radar nets.

The next day Kelly told us, "Well, the damned thing came out of China, but was lost. It wasn't spotted or shot down, but it must've malfunctioned and crashed on us."

Those monitoring the flight said Chinese radar never detected it, so we concluded the guidance had screwed up and the drone just kept on chugging until it ran out of gas, probably after crossing the Sino-Soviet border into Siberia.

About eleven months later, the Nixon White House again approved a flight against Lop Nor. This time the drone performed the mission perfectly, arriving back right on rendezvous point, but after dropping its camera and photo package, the chute failed to open and the package plummeted into the sea and was lost.

In March 1971, Nixon approved a third flight over the same area. This one also functioned perfectly; the drone separated its camera package on schedule and it parachuted into the sea where a Navy frigate was on station awaiting a pickup. Unfortunately, the seas were heavy and the Navy botched the recovery, allowing the package to be pulled beneath the ship with the parachute on one side and the hatch on the other. The package sank before they could get cables around it.

The final flight occurred two weeks later. This time our bird was tracked nineteen hundred miles into China and then disappeared from the screen. The reason was never determined. No other flights were attempted. The complex logistics surrounding each flight, involving recovery ships and rendezvous aircraft, cost a bloody fortune to stage. We were canceled in mid-1972.

And Kelly was bitter about it. "I'm not one to find scapegoats," he told me at the time, "but one reason why we had failures over China is that the birds we used had been stored up at Beale Air Force Base for nine months before the missions were authorized. The blue-suiters had 160 people there assigned to this program. Each of them had a salary to justify and took our drone apart frequently after our final checkouts here, just to put it back together again. And by so doing, they screwed up the works. We should have had the Skunk Works doing complete field service and even fly the actual missions and launch those birds. I'm telling you, Ben, that would have made all the difference in the world."

The end came in the form of a Defense Department telegram to Kelly on July 8, 1971, informing him that Tagboard had been canceled and ordering him to destroy all of its tooling. Still, we felt pride in the high degree of performance obtained for such a low cost.

On a February day fifteen years later, a CIA operative came to see me at the Skunk Works carrying a panel, which he plopped down on my desk. "Ben, do you recognize this?"

I grinned. "Sure I do. Where did you get it?"

The CIA guy laughed. "Believe it or not, I got it as a Christmas gift from a Soviet KGB agent. He told me this piece was found by a shepherd in Soviet Siberia."

Actually, it was from the first D-21 mission into China in November 1969, when the drone flew off course into Soviet Siberia before running out of fuel.

The panel was from the drone's engine mount. It was made from composite material loaded for radar absorption and looked as if it had been made just yesterday. The Russians mistakenly believed that this generation-old panel signified our current stealth technology. It was, in a way, a very nice tribute to our work on Tagboard.

13

THE SHIP THAT NEVER WAS

IN THE SPRING of 1978, while we were developing our model for the first stealth airplane, our project photographer stopped me in the hall to complain about defects in a new Polaroid camera we had recently purchased. "I've been taking instant view shots of the stealth model, and I'm getting very fuzzy pictures. I think I've got a defective lens," he remarked. I slapped my head, knowing we had accidentally stumbled onto an exciting development. "Time out! There isn't a damn thing wrong with your new camera," I insisted. "Polaroid uses a sound echo device like sonar to focus, and you are getting fuzzy pictures because our stealthy coatings and shaping on that model are interfering with the sound echo."

I was always on the prowl, looking for new ideas to expand or exploit technologies we were developing. A stealth airplane was our goal. But how about a stealthy submarine that would be undetectable on sonar? If we had avoided the sonar device built into a Polaroid camera, why couldn't we avoid sonar returns against submarines or even surface ships specially treated and shaped to escape detection? I had a couple of our engineers buy a small model submarine, put faceted fairings on it, and test it in a sonic chamber. Even with such a crude test setup, we discovered that we had reduced the sonar return from that model sub by three orders of magnitude. In the engineering game, improving anything by a single order of magnitude — ten times better — is a very big deal, usually worth a nice bonus or at least a bottle of champagne. Three orders of magnitude is *one thousand times* better. That's worth a fortune, a medal, or both, but is as rare an occurrence as an astronomer discovering a new constellation.

So we decided to design a stealthy sub; the cigar-shaped hull was shielded by an outer wall of flat, angular surfaces that would bounce sonar signals away and also muffle the engine sounds and the internal noises of crewmen inside the vessel. We ran numerous acoustical tests in special sound-measuring facilities and obtained dramatic improvements. If nothing else we had rendered null and void the favorite cliché of a lot of World War II naval action movies about a submariner sneezing or dropping a monkey wrench at the critical moment when the enemy destroyer's sonar search pings ever closer. The flat outer wall effectively eliminated any noises. Armed with high hopes, I took our design and test results to the Pentagon office of a Navy captain in charge of submarine R & D. By the time I left his office, I was grimly reciting Kelly's Skunk Works Rule Number Fifteen. Fourteen of his basic rules for operating a Skunk Works had been written out, but the fifteenth was known only by word of mouth, verbal wisdom passed on from one generation of employees to the next: "Starve before doing business with the damned Navy. They don't know what in hell they want and will drive you up a wall before they break either your heart or a more exposed part of your anatomy." I'd been a fool to ignore Kelly's wise words of warning.

That submarine captain epitomized the hidebound Navy at its worst. He frowned at my drawing and backhanded my concept. "We don't build submarines that look like that." He admitted that our test results were "interesting" but added, "Your design would probably cost us two or three knots in speed." I countered, "But why care about losing three knots, when you are invisible to your enemy?"

He ignored me. "This looks more like the *Monitor* or the

Merrimac from the Civil War," he said. "We'd never build a modern submarine that looked like *that*."

I returned to Burbank with a renewed healthy disdain for the anchors aweigh crowd, but one of our engineers, just back from a Pearl Harbor business trip, mentioned to me that he had seen a catamaran-type ship that the Navy had built experimentally on the q.t. out of unauthorized funds. This was a prototype SWATH (Small Water Area Twin Hull) ship that was proving to be amazingly stable in heavy seas and was considerably faster than a conventional ship. It seemed to me that a catamaran SWATH ship held real promise as a model for a stealthy ship. And on my next trip to Washington for a meeting on our stealth airplane design with Defense Undersecretary Bill Perry, who was the Carter administration's czar of stealth, I mentioned the idea of a model stealth ship. I told Dr. Perry that the catamaran would provide a perfect test of the effects of stealth shaping and coatings for surface vessels. We also wanted to test the effects of seawater on radar-absorbing iron ferrite coatings. Dr. Perry agreed and ordered the Defense Advanced Research Projects Agency (DARPA) to authorize a study contract with us.

I put our best special projects engineer, Ugo Coty, in charge. DARPA had come up with $100,000, and I kicked in an additional $150,000 to begin developing a workable model catamaran. One of the biggest threats against our surface vessels was the Soviet RORS satellite, using powerful X-band radar. Shape was the key to defeating Soviet radar. Coatings accounted for only 10 percent effectiveness in deflecting radar. The rest was quietness of a vessel's engines and minimizing its wake.

Ugo picked four other engineers for his team and set to work developing a SWATH ship model that sat on the water like a catamaran with a pair of underwater pontoon-type hulls that propelled the ship with twin screws. The underwater pontoons provided most of its buoyancy and good stability in rolling seas and also produced very little wake. A ship's wake is as easy to spot from the sky as a fighter's contrail.

Our ship was the most unconventional seagoing vessel ever to come off a drawing board. There was a definite family resemblance to our stealth fighter. Only the floating version had no wings. It was a series of severe flat planes at 45-degree angles that sat above the water on struts connected to a pair of submerged pontoons. The ship would be powered by the diesel-electric propulsion that drives electric generators. Cables carried the current to a pair of powerful electric motors in each submerged pontoon that spun counter-rotating propellers. Careful shaping of the pontoons and the propellers cut down sharply on noise and wake.

This wasn't exactly a ship of classic design, but my hunch was that we could really fill an important niche in the Navy's defensive needs. NATO war games played out by computer had triggered alarm among the Navy brass about the vulnerability of our carrier task forces to enemy air attack. The premise of the game was that the Soviet Union had attacked Western Europe and that the U.S. Navy had quickly reacted by steaming a backup carrier task force into the North Atlantic.

Unfortunately, Soviet long-range fighter-bombers using new look-down, shoot-down radar-guided missiles caused worrisome losses to our carriers and escorts in the computerized warfare exercise. To counter this threat, the Navy was crashing production of a billion-dollar missile frigate that would fire the new Aegis ground-to-air missiles designed to destroy incoming cruise missiles. I thought, Why go after the arrows? Go after the shooter. To the chagrin of the billion-buck Aegis frigate backers, our SWATH boat would cost only $200 million. We could arm it with sixty-four Patriot-type missiles and send it out three hundred miles ahead of the carrier task force as an invisible, amphibious SAM missile site. We'd shoot down the Soviet attack aircraft before they got in missile range of the fleet. And because they couldn't see the stealth ship electronically, they'd literally never know what hit them. We did an analysis and determined that the entire U.S. Navy carrier fleet could be protected by only eighteen of our stealth defenders armed with SAMs. Since our ship would knock out most of the incoming air armada trying to attack our carriers, we would make the Aegis more effective by dramatically decreasing the number of incoming missiles it would be called upon to try to destroy at one time.

And we had compelling test data to prove our contentions. To accurately measure the low radar cross section of a model ship under realistic ocean conditions, Ugo Coty went to one of the most remote areas of the western desert. There, within hailing distance of Death Valley, he built himself a miniature ocean. The place was about two days from Edwards Air Force Base by mule pack or by jeep crawling through deeply rutted roads. It was an ancient lake bed sitting at the foot of an 8,500-foot mountain, on top of which Ugo planned to mount a radar system that would duplicate the Soviet radar satellite's. Ugo's ocean was a hundred-foot-by-eighty-foot plastic swimming pool eight inches deep. The model sat atop a thirty-foot table that rocked and rolled to approximate a realistic ocean. The Soviet's X-band radar was hauled up the steep side of the mountain in a secondhand refrigerated meat truck that kept both the electronics and the operators from frying in the broiling sun. To get up the steep dirt incline, the truck had to be pushed by a bulldozer. The biggest problem we encountered was with the region's wild horses and mules, whose ancestors were the pack animals used by borax miners who dug in those mountains during the last century. The smell of water drew the wild herds to our make-believe ocean and interfered with our testing until one of our guys solved the problem by adding buckets of salt to our bogus ocean. Soviet spy satellites overflew once a day because the Navy used that desert range for missile testing, and the Russians must have wondered what in hell we were up to out there, stocking a pool near Death Valley for thirsty mules and horses.

In the early fall of 1978, I took our test results to Bill Perry at the Pentagon. I reviewed with him all our tests and the low radar returns we had managed to achieve so far. He was enthusiastic and ordered the Navy to provide research and development funding for the creation of our prototype stealth ship. The ship would be called *Sea Shadow*. But I returned to Burbank with some anxieties about the mixed blessings of launching another stealth project while we were still struggling to build the first stealth airplane, were currently updating our fleet of U-2s, and producing SR-71 Blackbirds. I was thriving while our shipbuilding people in Northern California were laying off workers. That situation was not lost on the Ocean Division executives, who heard rumors about our secret stealth ship project and complained to the Lockheed top management that they needed this project one hell of a lot more than did the Skunk Works. I found myself suddenly under pressure to let go of the ship and send it northward.

God, I hated to do it, but company politics and basic management considerations prompted me to go along and give up the ship project. I had a full plate without it, while my shipbuilding colleagues were scrambling to find business. I surrendered to the inevitable and turned over the project to the Ocean Division, but only after I convinced our CEO, Roy Anderson, to allow Ugo Coty and his team, who had done all the work on the ship in the test phase, to stay on as overseers of the actual construction. "We need Ugo to keep those damned shipbuilders from going off on a tangent," I told Roy Anderson. "This is one project where the method of shipbuilding is much less important than the stealth technology," I told Roy. "They'll want to sacrifice the stealth if it gets in the way of the ship's performance, but Ugo will force them to stay focused. All Dr. Perry wants to prove out is the stealth. That's key to this test. If the ship merely floats that's good enough." As it happened, my fears about the conflicting agendas between professional shipbuilders and experts on stealth technology, like Dr. Perry, were realized almost from the first day that the Ocean Division took over the project.

Ugo Coty did his best, but he ran into heavy weather. His original six-man operation quickly was shunted aside by eighty-five bureaucrats and paper-pushers running the program for the division. Then the Navy marched in, adding its supervision and bureaucracy into the mix with a fifty-man team of overseers, who stood around or sat around creating reams of unread paperwork. No ship ever went to sea — not even a top-secret prototype — without intensive naval supervision to ensure that all ironclad naval rules and regulations were strictly enforced before the keel was ever laid.

"Where is the paint locker?" a Navy commander demanded of Ugo, rattling the blueprint plans. Since the days of John Paul Jones, every naval ship afloat has a damned paint locker on board. *Sea Shadow* would definitely not be the only exception since the Revolutionary War, and before she started her sea trials, a paint locker was located just below the bridge. Maintaining secrecy was another big headache. How do you build a 160-foot ship, whose strange, exotic appearance and shape are the keys to its secrecy, and keep that shape and appearance from the eyes of a small army of shipyard workers building it? The only logical answer was to build the ship in pieces, doling out six or eight sections per shipyard, then assembling the pieces as a gigantic jigsaw puzzle inside a huge submergible barge located in Redwood City, California.

Sea Shadow was made of very strong welded steel, displaced 560 tons, and was 70 feet wide. During early sea trials in 1981, we suffered unexpectedly large wakes that were easy to spot on radar and from the air, which completely baffled us until we discovered that the motor propellers had been installed backward! Our ship had a four-man crew — commander, helmsman, navigator, and engineer. By contrast, a frigate doing a similar job had more than three hundred crewmen.

Viewed from head-on the ship looked like Darth Vadar's helmet. Some Navy brass who saw her clenched their teeth in disgust at the sight of the most futuristic ship ever to ply the seas. A future commander resented having only a four-man crew to boss around on a ship that was so secret that the Navy could not even admit it existed. Our stealth ship might be able to blast out of the sky a sizable Soviet attack force, but in terms of an officer's future status and promotion prospects, it was about as glamorous as commanding a tugboat. At the highest levels, the Navy brass was equally unenthusiastic about the small number of stealth ships they would need to defend carrier task forces. Too few to do anyone's career much good in terms of power or prestige. The carrier task force people didn't like the stealth ship because it reminded everyone how vulnerable their hulking ships really were.

But by the fall of 1982, when Britain and Argentina began hostilities around the Falkland Islands, the *Sea Shadow* won renewed attention at the Pentagon. The British reported success using primitive ferrite-coated nets hung over the masts and radar antennas of their warships to lower their radar cross section against air attack. Even the *QE 2*, which was drafted as a troop transport, used these antiradar nets. The Brits claimed to have lowered their radar cross section by an order of magnitude. But their results weren't one quarter as impressive as ours. A typical warship was a very high reflector of radar — a radar profile equal to about fifty barns. Our frigate would show up a hell of a lot smaller than a dinghy.

By the time we were ready for full-scale testing in the early summer of 1985, the Navy was eager to subject our prototype to the most rigorous radar testing imaginable. Several of their radar experts claimed that there was no way we could duplicate the low radar cross section achieved by a thirty-foot model in a pool with a full-size prototype on the real ocean. We had heard those same skeptical predictions before from the Air Force over our stealth airplane, but in the wonderful world of stealth, once we had acquired the right shape, the size of an object really didn't matter. The military had a tough time understanding that basic fact.

The barge, with *Sea Shadow* cocooned inside for secrecy, was towed from Sunnyvale in Northern California down to Long Beach to begin its tests in the dead of night, off Santa Cruz Island, where we had constructed a radar installation. We sailed our ship against the most advanced Navy hunter planes. To make certain that their radar detectors were functioning by the book, we placed submarine periscopes in the ocean at the nautical mile range where the plane's onboard radar would first be expected to detect them. The radar worked predictably, picking up the radar signature of the periscope at the expected range. Then the airplanes flew on, seeking our ship. On one typical night of testing, the Navy sub-hunter airplanes made fifty-seven passes at us and detected the ship only twice — both times at a mile-and-a-half distance, so that we would have shot them down easily long before they spotted us. Several times, we actually provided the exact location to the pilots and they still could not pick us up on their radar.

These kinds of tests went on for nearly a year and often were conducted under the scrutiny of Soviet trawlers snooping in the open seas off the Channel Islands, about sixty miles southwest of Santa Barbara. To keep local fishing boats and any curious yachtsmen away, the Coast Guard leaked the word that they were escalating stop-and-search procedures against potential drug smugglers in the ship lanes we were using. Boat traffic miraculously vanished.

The tests began an hour after sunset and lasted until an hour before sunrise. Then *Sea Shadow* was docked inside the barge and tugged back to Long Beach. But the long, strenuous months of testing confirmed most of our original stealth predictions. One of the biggest problems we had to overcome was our own extreme invisibility! The ocean waves showed up on radar like a string of tracer bullets. And if the ship was totally invisible, it looked like a blank spot — like a hole in the doughnut — that was a dead giveaway. In the stealth business, you tried like the devil not to be quieter than the background noise, because that was like a trumpet-blast warning to the enemy.

By the time we solved this problem, however, the admirals who ran the surface fleet were displaying little enthusiasm for going any speed ahead. "Too radical a design," they told me. "If the shape is so revolutionary and secret, how could we ever use it without hundreds of sailors seeing it? It's just too far out." There were sexier ways of spending naval appropriations than on a small secret ship that would win few political brownie points for any admiral who pushed for it. Although the Navy did apply our technology to lower the cross section of submarine periscopes and reduce the radar cross section of their new class of destroyers, we were drydocked before we had really got launched. So I held back: I had a design for a stealthy aircraft carrier that would show up on radar no bigger than a life raft, but I had already proven Kelly's unwritten Rule Fifteen about dealing with the Navy. Why ignore it twice?

14

THE LONG GOODBYE

IN THE EARLY SUMMER of 1972, an executive of Northrop Aircraft Company, whom I'll call Fred Lawrence, invited me to dinner and offered me a terrific job.

"Ben," he said, "we are planning to build a lightweight single-engine fighter. We've never done a single-engine jet before, and we want you to set up a Skunk Works for us with you as our Kelly Johnson."

"Why me?" I asked, truly amazed. I could think of three or four veterans of our Skunk Works who had a world of managerial and practical experience greater than my own. For openers, our vice president, Rus Daniell, had been a project manager before I was even hired at Lockheed, and on the basis of seniority would certainly appear to be more qualified than I was. I was then forty-seven years old, working as Kelly's assistant chief engineer, in charge of aerodynamics and all the flight sciences. I had about forty-five people working for me and was earning sixty grand. I was doing dandy, without any real future aspirations.

"Why you?" Fred Lawrence smiled, repeating my question. "Because you've got the temperament and background we've been looking for. You're Kelly's troubleshooter on the technical side, which is recommendation enough. We need you to make sure we get the right propulsion system to fit performance capability. That's eighty percent of the battle — am I right? How many prototypes go down the toilet because the

engine is underpowered or totally wrong for what the designers want to achieve? You won't let that happen. You have a
solid reputation for being a straight shooter and customers
like you. Ben, with you on board we think we can do big
things in the fighter market."

He explained that they wanted to go after the cheaper,
lighter airplanes for the NATO and Asian markets, and had
plans in the works to prototype an advanced Air Force interceptor. "We want a piece of that fighter market," he said, "and
you can help us grab it. You know that won't happen at the
Skunk Works as long as Kelly is in charge. The Air Defense
Command crosses the street when they see him coming. He's
just too big a pain in the ass to work with. You know I'm telling it right. Ben, break out and come with us. Be your own
man. I'm telling you, you won't regret it. We'll do big things
together."

I didn't trust Lawrence, but the job he was offering represented the kind of personal challenge that I lived for. Over
coffee, he told me that whatever salary Kelly was currently
paying me, Northrop was willing to better it by ten grand annually. I promised them a decision in a week.

My wife, Faye, was no help. She said, "Ben, only you can
decide what to do. It is up to you." The truth was that I was
tempted to grab the job, but dreaded having to confront Kelly
with my decision.

In the past six years he had taken me under his wing, putting me up for various professional awards for my work on
the Blackbird's revolutionary moving-spike inlets, encourag-
ing me to write technical papers for aeronautical journals to
increase my name recognition within the industry. The curse
of operating inside a top secret world is that very few in the
aerospace industry knew you even existed. I also had talked
Kelly into letting me attend a thirteen-week management
training course at Harvard's Business School in the summer
of 1969. It was an advanced management institute for about

one hundred and fifty carefully selected, upwardly mobile
executives — and Kelly wrote me a glowing recommendation
that helped me get in, and authorized Lockheed to pay the
tuition freight, which was considerable. He backed me even
though he insisted that it would be a complete waste of my
time. "I'll teach you all you need to know about running a
company in one afternoon, and we'll both go home early to
boot. You don't need Harvard to teach you that it's more important to listen than to talk. You can get straight A's from all
your Harvard profs, but you'll never make the grade unless
you are decisive: even a timely wrong decision is better than
no decision. The final thing you'll need to know is don't half-
heartedly wound problems — kill them dead. That's all there
is to it. Now you can run this goddam place. Now, go on home
and pour yourself a drink."

But I had persisted, and when I returned from Cambridge,
wearing a new crimson tie, Kelly asked me for my appraisal of
the Harvard Business School. To accommodate him, I wrote
out an equation: $\frac{2}{3}$ of HBS = BS. He roared with laughter, had
my equation framed, and gave it back to me for Christmas.

Kelly let me test my Harvard training by putting me in
charge of an elite Skunk Works team of engineers and designers that he loaned to Lockheed's main plant for six
months, just down the street from our Burbank headquarters,
in the fall of 1969, to help them build five prototypes for a new
carrier-based Navy submarine-hunting airplane. He could
spare us because our own business was unusually slack. Between the new Blackbird and the older U-2, the Air Force had
all the spy planes it could use. And that was our house specialty. Northrop's Fred Lawrence was correct: the blue-suiters
who controlled fighter procurements didn't think of us as
fighter builders any more. Our last fighter was the Mach 2
F-104 Starfighter, built during the Korean War to match the
fastest Russian MiGs. But for more than a decade we had
been seldom included in the bids sent out from the Pentagon

announcing a competition for a new fighter. Lockheed as a corporation had a booming business in cargo aircraft and missiles for the Navy. But the Skunk Works had to lay off about eleven hundred workers. And the open secret, which several of us inside the Skunk Works realized but never openly discussed out of loyalty to the boss, was that the blue-suiters in charge of fighters had blackballed Kelly Johnson and excluded him and us from fighter competition because they found him too contentious and bullying to work with any longer.

So I really felt rotten when I finally worked up the courage to inform Kelly about Northrop's offer to start up a fighter project. In his dealings with me, he appreciated the fact that I always provided him with alternatives when presenting him with a problem: I'd say, "Solution one will cost you so much money. Solution two will cost you so much time. You're the boss. Which do you choose?" So often, others would come to him like errant schoolboys and moan, "Kelly, bad news. We broke that part." And he'd get sore and shout, "Well, what in hell do you expect me to do about it?" Now I had no alternative to offer for the course of action I was taking. I was walking out on him.

During the time I spent at the main plant on the Navy sub-hunter project, Rus Daniell had to suddenly step into the breach and take over in place of Kelly, who was hospitalized with a serious abdominal infection. That brief stint running the Skunk Works for a couple of months in 1970 proved to be a personal disaster for Rus. He had signed off on a project proposal for the Air Force that included a glaring mathematical error. The Pentagon analysts who discovered the mistake couldn't resist rubbing our noses in it, and a two-star general, who had probably waited for years to stick it to Kelly and his know-it-alls in Burbank, phoned him in his hospital room and raised hell about our sloppy work. Kelly was livid and ordered me back to the Skunk Works immediately, to take charge of the technical section. Rus Daniell had to obtain my approval on all technical matters. Since he was a vice president and the odds-on favorite to be Kelly's successor, I'm sure he wasn't delighted with that arrangement, but he handled it gracefully. From then on, when there was a critical meeting about deadlines or problems or an angry customer in Kelly's office, I would be sent for to sit in, sometimes with Rus, but more often without him.

Kelly and I had grown close. He had lost his wife Althea in the fall of 1970, after a long struggle with cancer. She was his age and had been his secretary years earlier. Before she died, Althea told Kelly that he needed to remarry without much delay because he was not the kind of man who could live alone. She worried about his drinking and poor eating habits when left to himself. "I think you should marry MaryEllen," Althea told him. MaryEllen Meade was Kelly's secretary. She was a vivacious redhead, twenty-five years younger than both of the Johnsons, and had recently suffered a messy divorce. She had started her Skunk Works career several years earlier as my secretary, and both my wife and I got to know her well and were fond of her. She was in awe of Kelly and very devoted to him.

About six weeks after Althea's funeral, Kelly sent for me. He seemed embarrassed and troubled, and I knew he was in a real quandary when he began confiding his personal problems, something very much out of character. But he told me about Althea's deathbed wish that he marry MaryEllen and wondered what people would think if he carried out that wish any time soon. He fretted over the age difference and how awkward it would be if MaryEllen refused him.

My heart went out to him. "Kelly," I said, "since when do you worry about what people think? All that matters is what *you* think. No one around here will think you're a dirty old man, if that's what you're worrying about. We'll all be secretly jealous as hell. I don't have the slightest doubt that MaryEllen could make you very happy." He thanked me and proposed to MaryEllen a few days later. When the couple returned from their Hawaiian honeymoon, in June 1970, he began inviting Faye and me to join them for dinner or to be weekend guests at Kelly's 1,200-acre cattle ranch, near Ronald Reagan's big spread, overlooking Solvang, California. Kelly had seldom socialized with any employees except at occasional company functions, or even more rarely, asking a couple of guys to join him for a quick round of golf at a local country club, where Kelly would breeze through eighteen holes using only a six iron, which he claimed was just fine for all shots, near and far. He was so muscle-bound from having been a hod carrier as a kid that his golf swing for distance was quite limited. But he was a skilled putter until occasional swigs from his back-pocket flask got the best of his aim.

I was not a golfer, but I adored Chinese food and when Kelly discovered we had a mutual passion, my wife and I began joining him and his new bride for weekly Chinese culinary outings at favorite restaurants. Kelly seemed to be genuinely happy with MaryEllen and eager to show her off. She even got Kelly to give up drinking scotch because she hated the smell on his breath; he began drinking vodka instead. Faye and I would be invited to accompany the newlyweds on the Lockheed jet on weekend trips to the Air Force Academy in Boulder, Colorado, or to aeronautical conventions in other picturesque places where he gave a speech and then introduced his very attractive new wife with obvious pride and affection. "MaryEllen loves Faye," he confided to me, "and I think it's about time that the right people in this business get to meet you too."

But just around the time that Northrop approached me in 1972 on the new job, MaryEllen began getting sick, seriously sick. She was a diabetic and didn't really take care of herself. Her diabetes became a monster unleashed, and she never was able to catch up with it. She developed kidney problems, then her eyesight began to fail. Kelly was heartsick, but outside of myself and Norm Nelson, no one else at work was aware of the extent of MaryEllen's illness. When she learned of Faye's sudden death in late 1980, she was devastated. By then she weighed about eighty pounds. She said to me, "Oh, Ben, Faye was so much stronger and healthier than I. What hope can there be for me?" She died a few weeks later. She was only thirty-eight years old.

Before she died she advised Kelly to marry her best friend, Nancy Horrigan — which he did, soon after MaryEllen's funeral, and after again consulting with me about what people might think, which was probably just another way for him to discover what *I* thought. I told him, "Kelly, do it. You need a good woman in your life. MaryEllen had only your best interests at heart."

I told Kelly about Northrop's offer on a day when he was leaving early to go with MaryEllen to the hospital for tests that would ultimately reveal her need for a kidney transplant. When I discovered how unfortunate my timing was, I felt dreadful. I could tell he was worried and preoccupied, but I had no choice but to push ahead and blurt it out. "Kelly," I said, "you've been like a father to me. I love it here. I love the work and the people and the uniqueness of this place. But this is a golden opportunity."

I laid out Northrop's offer, and he closed his eyes and solemnly shook his head. "Goddam it, Ben, I don't believe a word that guy said to you. I'll bet my ranch against Northrop starting its own Skunk Works. Companies give it lip service because we've been so successful running ours. The bottom line is that most managements don't trust the idea of an independent operation, where they hardly know what in hell is going on and are kept in the dark because of security. Don't kid yourself, a few among our own people resent the hell out of me and our independence. And even those in aerospace who respect our work know damned well that the fewer people working on

a project, the less profit from big government contracts and
cost overruns. And keeping things small cuts down on raises
and promotions. Hell, in the main plant they give raises on the
basis of the more people being supervised; I give raises to the
guy who supervises least. That means he's doing more and
taking more responsibility. But most executives don't think
like that at all. Northrop's senior guys are no different from all
of the rest in this business: they're all empire builders, because that's how they've been trained and conditioned. Those
guys are all experts at covering their asses by taking votes on
what to do next. They'll never sit still for a secret operation
that cuts them out entirely. Control is the name of the game
and if a Skunk Works really operates right, control is exactly
what they won't get."

He smiled and rubbed his eyes. "Use your head. Ben, believe me, they just want you over there to pick your brains on
getting that fighter prototype going right, then use you as the
wedge to try to steal more of my best employees. But mark my
words, you'll be reporting to a dozen management types and
they won't let you out of their sight for one minute."

He stood up from his desk and came around to my chair
and looked down at me. "Ben, forget Northrop. This is where
you belong. I don't want to lose you. So let me put my cards on
the table. Would you stay put if I matched Northrop's offer?"

I was stunned. "But, Kelly, you already have a vice president. What about Rus Daniell?"

"Who says I can't have two veepees?" he replied, patting me
on the shoulder. He said he would name me vice president for
advanced projects, just as Northrop proposed to do, and make
Rus Daniell his vice president for current projects. "You're the
guy who has the vision and the ideas," he said. "You're the guy
who can make things happen down the road. I count on your
imagination." Then, for the first time in his dealings with me,
he discussed his retirement in three years' time and told me
that I was his personal choice to succeed him. "I've got to give

the board a choice when I hit sixty-five. It will be between Rus and you. But he's too old, only two years younger than I am. I will make my wishes known to the board and that won't exactly hurt your chances. I've had my eye on you for a long time. You've got the brains and personality to do the job. We've got a lot of talented engineers around here, but not too many natural leaders. I like the way you get along, how you deal up-front with everyone, with your good spirit and your energy. So, goddam it, between now and my retirement party, don't you dare to screw up."

He also matched Northrop's offer of a ten-thousand-dollar raise. It was a counteroffer I joyfully accepted.

He began taking me, in my new role as advance planner, along on his trips to the Pentagon. All of us at the Skunk Works knew the two basic rules for getting along with Kelly Johnson: all the airplanes we built were Kelly's airplanes. Whatever pride we secretly took, we kept to ourselves. And if a blue-suiter wore a star on his shoulder, only Kelly Johnson was authorized to deal with him. The rest of us were free to establish relations with bird colonels and other underlings. Of course, by the time that Kelly retired, many of those colonels I was cultivating would become generals and take command, while Kelly's connections would be shuffling off to enjoy Boca Raton or Palm Springs on their government pensions.

After all, he had been a familiar figure around official Washington since World War II. He had built the Hudson bomber for the Brits, the P-38 fighter and the P-80 and the F-104 — all his recommendations for those projects had been enthusiastically accepted without the slightest argument, the brass showing him the greatest deference and respect. Most of them were former pilots who couldn't read a blueprint or change a spark plug, much less match wits or dare to contradict the great Kelly Johnson, the genius with the slide rule who created some of America's greatest flying machines. So Kelly told them what they ought to be doing and they saluted smartly.

Kelly loved to tell how a general named Frank Carroll was so enthusiastic hearing Kelly describe the speed and maneuverability of the new P-80, America's first jet, which he had been pushing for, that Carroll decided to bypass all the red tape delays and do all the purchase order paperwork himself. "We came back from a quick lunch at two in the afternoon. He had an official letter of intent for me to start work on the P-80 drafted, approved, signed, and sealed in time for me to catch the 3:30 flight back to California," Kelly said, chuckling delightedly every time he told that story. The same thing happened with the F-104 Starfighter. General Bruce Holloway, who was then head of SAC, was a colonel in procurement back in the 1950s, and listened to Kelly's pitch about building a supersonic jet. Holloway needed to obtain a list of Air Force requirements to match Kelly's performance description as the first step toward forwarding a contract for a prototype. "By God, Kelly, I'll write it myself," he declared in a blaze of enthusiasm. Kelly helped him draft it, and the two of them carried it up the chain of command to a general named Don Yates, who signed off on it. Total elapsed time: two hours.

Those days were gone forever. Now it was the Air Force calling all the shots, and a towering figure like Kelly Johnson, with his two Collier Trophies and his presidential Medal of Freedom, was respected for his past accomplishments by brash, young aeronautical engineer graduates of the Air Force Academy who had little interest in most of the ideas he tried to generate for new airplanes. Ideas were now a one-way street, initiated by Air Force planners with doctorates in flight sciences. It was rare that the Air Force took a manufacturer's idea and ran with it, rarer still that manufacturers bothered to present unsolicited proposals to the Pentagon's planners. And even Kelly was forced to admit that aeronautical brainpower was no longer our monopoly: several of those young procurement officers we dealt with were sharp enough to be hired at the Skunk Works.

Kelly's stature still gained him entry to the offices of the secretary of the Air Force, the Air Force chief of staff, the head of the Air Force's black, or top secret, programs, the director of the Central Intelligence Agency, Senator Goldwater, and half a dozen other of the top movers and shakers. Through him, I met them all. But Kelly being Kelly had also made as many high-level enemies as friends, and plenty of blue-suiters who had dealt with him on procurement matters still had the wounds to prove it. Kelly said what he meant and meant what he said, and he couldn't care less about who was offended. I remember attending one black-tie affair with him in late 1974 at which Barry Goldwater, a general in the Air Force Reserve, was being honored with an award, and half the Air Force high command was present. Kelly was asked to make a few remarks. He said: "I've never got in trouble making a speech on engineering. But over the years I've learned what not to say. I'm an expert at that. For instance, I will not say that if we had followed the policies of the man we are honoring tonight we would not have made such a shambles of the Vietnam War as we did. Nor would I say how stupid we were in the same war to take the guns out of our fighter aircraft and then send our pilots into combat with such a cost-saving training program that few of them had fired even a handful of air-to-air missiles before facing the enemy. And for gosh sakes, I won't say that the Israeli air force using the same kind of U.S. aircraft against the same kind of MiGs scored about four times the victory ratios that we did, using better pilot training and tactics!"

When he was finished speaking, about half of those blue-suiters present sat on their hands. Still, we had our shots at getting back in the fighter business. In 1972, the Air Force put out bids for a new lightweight fighter to several companies. We managed to insinuate ourselves into the bidding. This would be an advanced version of the F-4 Phantom, a highly maneuverable Air Force tactical aircraft that NATO allies could use as well to upgrade their air defense inventories. The bid contained specific design and performance requirements for a fighter that weighed 17,000 pounds, carried 5,000 pounds of fuel, had a 275-square-foot wing, and a specific excess power rate. I was ready to put my design team on it and give the Air Force exactly what they asked for, but Kelly had other ideas.

"Goddam it, Ben, this airplane isn't carrying enough fuel. A fighter on afterburner uses a thousand pounds a minute, and every fighter that carried only five thousand pounds in Vietnam combat ran out in tough spots and the pilot wound up as a guest at the Hanoi Hilton. This is unacceptable. I won't submit a proposal for something this wrong."

I pleaded with Kelly to go along. I said, "Give the Air Force what they ask for initially, and as the procurement process unfolds all kinds of changes and modifications will follow. That always happens; you know that. But if you cut us off before we even get up to the plate, there's no way we can score."

He got sore. "Ben, if I teach you anything, it's this: don't build an airplane you don't believe in. Don't prostitute yourself for bucks."

In the end, he allowed us to submit his version of what he thought the Air Force should be requesting: a fighter weighing 19,000 pounds and carrying 9,000 pounds of fuel and a 310-square-foot wing. General Dynamics won the competition by sticking to exactly what the Air Force wanted. By the time their airplane, the F-16, became operational, it weighed 19,000 pounds, carried 7,400 pounds of fuel, and had a 310-square-foot wing. I told Kelly, "Admit it, the blue-suiters would have made your changes in due course, just as I predicted."

But they weren't about to be led by the nose by Kelly Johnson. And to compound the loss, which was substantial, because General Dynamics sold hundreds of those damned fighters, Kelly had balked over the Air Force's insistence that they run all the flight tests on the prototype of the fighter, if we had been selected. "I will never agree to that," he fumed. Historically, we flight-tested our prototypes and made all the fixes and adjustments before turning them over to the customer. Kelly called Lt. General Jim Stewart, head of the Aeronautical Systems Division at Wright Patterson Air Force Base in Dayton, Ohio, and read him the riot act. "No one runs my flight tests, Jim, but me and my people. That's the way it's been since day one. And that's how it is going to stay. I'm putting you on notice right here and now."

After that phone call, I received a message at home from General Stewart's administrative assistant: Stewart was fed up dealing with Kelly Johnson, and so was the fighter command. As far as Tactical Air Command was concerned, the Skunk Works could go fly a bomber. We were out of the fighter business.

Kelly's stubbornness angered several important bluesuiters and a few of our own senior corporate executives. And in the fall of 1974, as his sixty-fifth birthday approached, he sent for me and informed me that I was his personal choice to succeed him at the Skunk Works, and he had so informed the company's president, Carl Kotchian. "I told him that, out of fairness, I had to submit Rus Daniell's name too, but that you are the one I've trained and counted on to carry on the great Skunk Works tradition." He looked tired and discouraged, but also I detected a sense of relief. And a few days later Kotchian called me and I began jumping through the hoops at a series of meetings with key senior executives. All of them demanded the same thing: that I accommodate our blue-suit customers and that I recognize management's responsibilities to keep a close eye on me as Kelly's successor until I had a chance to prove myself. The name of the game was "get along and go along" — no more tyrannical Kelly Johnson types. I understood, as did Kelly, that he was unique in his power and independence, which was nontransferable to any successor. For better or worse, a new era was dawning for those left behind in a Kelly Johnson–less world. The Skunk Works was still expected to produce giant results, but the new guy sitting in the boss's chair would be a lot smaller than the original Gulliver.

Other Voices

Roy Anderson
(Former CEO, Lockheed Corporation)

Around the time Kelly chose Ben Rich to replace him, I was Lockheed's chief financial officer, dealing with twenty-six banks who were very nervous about the company's future in the aftermath of the scandals and the drubbing we took financially with the L-1011 airliner. But I was close to Kelly and knew Ben well, so Carl Kotchian called me and asked me what I thought about Rich as Kelly's replacement. The alternative was either Rus Daniell or hiring an outsider. I said, "I'd go with Ben. He's been handpicked and trained by Kelly. He's innovative and has terrific drive and energy." Carl said that was his inclination too. He said that Kelly had told him, "I've given Ben the toughest assignments and he's never let me down. He won't let you down either. He's the future. Rus is a good man, too. But Ben is better for this job."

Rus Daniell would have been first choice for a traditional kind of management. He was smooth and had a thoughtful, introverted quality that often inspired trust. But Ben Rich was a Skunk Worker through and through. He was an extrovert, high-intensity, no B.S. kind of guy. He told outrageous jokes and talked faster than a machine gun when he got excited about something, which was most of the time. He was just like Kelly when it came to problem solving and pushing things ahead — they were a couple of terriers who never let go or gave up. Kelly called me late at home one night and personally lobbied me about Ben. He had a couple of belts in him and he said, "Goddam it, Roy, I raised Ben in my own image. He loves the cutting edge as much as I do, but he knows the value of a buck and he's as practical as a goddam screwdriver. He'll do great, Roy. Mark my words."

Ben was already well known and respected in places that counted inside the CIA and among the blue-suiters. He had an instinctive sense of what came next in technology that was valued by the military and the agency. He was much more collegial than Kelly, more willing to compromise and stay loose in dealing with difficult customers. Ben also was a good politician: he recognized management's responsibilities and kept those of us who were cleared to know informed on projects and brought us in on big decisions. Kelly's attitude had been "Goddam it, if I tell you something, that's it."

In later years, after I became Lockheed's CEO, whenever I felt down in the dumps I'd call up Ben and drop by the Skunk Works. And I always left feeling a hell of a lot better. What cheered me up was Ben's enthusiasm, which he instilled in everyone else. Those guys, from engineers to shop workers, stayed focused. They worried about being on time, getting it right, and staying on budget. You just didn't find that kind of attitude anywhere else at Lockheed or any other company in the industry. That was the essence of the Skunk Works, and the reason why Ben would come into my office so many times over the years, with a big grin, saying, "Guess what our profits will be this next quarter."

After the F-16 loss, it took breakthrough stealth technology to get us back into the fighter business in the mid-1980s. By then, of course, I was in charge of the Skunk Works while Kelly became an increasingly uninvolved consultant. Watching me operate from his own desk, he kept his second guesses to a minimum, but I could usually tell when he disapproved of something I was doing. I said to him, "Kelly, I know what you're thinking, but it's a different climate now. The trick is to make the customer think he thought up the changes that we want. That's the easiest way to get these changes through. But, Kelly, that's a trick you never had the patience for." He had to laugh. "I'd never have let any of those dumb bastards second-guess me," he agreed.

But, unlike me, he never had to put up with an aggressive Air Force that challenged our Skunk Works autonomy at every turn. Because of highly publicized cost overruns that were suddenly endemic in aerospace, as well as headline accusations of bribes and scandals infecting many of the major defense companies, even we in the Skunk Works were now swarming with auditors and inspectors. We had maybe six auditors on the Blackbird project, now there were thirty of them on Have Blue, each one sniffing for evidence of grand larceny at every turn. With auditors came inspectors and security supervisors, who poked around in our waste bins and desks after hours looking for rule-breakers who didn't lock up or left secret work papers on their desks. They were even limiting the number of people I could clear for security for any one project. Each project had a specific quota of secret clearances allocated by the Pentagon, and each clearance had to be justified personally by me and approved by the Air Force. And Kelly, who was coming in only once a week to consult with me, was tough to justify with such tightly allocated clearance slots. "What's going on?" he'd ask, and it almost killed me because I couldn't tell him. He was no longer in the position of need to know.

But we still shared our own secrets. In July 1982, I confided to him that after two years of widowerhood, I was now in love again. I found a wonderful woman named Hilda Elliot, and planned to marry. Hilda was the manager of an antique shop and as smart and personable as she was attractive. She had three wonderful children from a previous marriage, and over the months of our dating game she had proven to be a great sport in putting up with industry functions where the technical talk would make any outsider's eyes glaze over. Kelly and his third wife, Nancy, quickly embraced Hilda, and she, in turn, became close to both of them over the next eventful years. Kelly, in particular, loved the story of Hilda's first trip with me to the Paris Air Show in 1983. The Russians invited us to visit their new C-5-like cargo plane, and when the CIA heard about it, they asked me if they could have one of their own technical experts accompany me while posing as just a personal friend I dragged along. The expert turned out to be a young and attractive brunette, and Hilda walked with her arm-in-arm on board the Soviet C-5 and introduced her as her cousin.

In 1986, Kelly broke his hip in a nasty fall and had to go to the hospital. He was admitted to St. Joseph's Hospital, only a few miles from our offices in downtown Burbank. And he never came out. He died in his hospital suite four years later. The problem was not his hip, which slowly mended, but a general physical deterioration and advancing senility exacerbated by hardening of the arteries to his brain. He died slowly and terribly. Hilda and I were frequent visitors to his hospital room at first, but as the months passed Kelly's condition deteriorated to the point where he stopped eating and his robust two-hundred-pound frame dissolved into a one-hundred-and-thirty-pound skeleton. His skin became a white and dry parchment and he suffered from excruciating bedsores. His eyes seemed unfocused and lifeless, and he increasingly began to slip in and out of coherence. I could barely stand to visit him, and many times he seemed not even to recognize me. But I came as much for Nancy's sake as for his. He wanted her at his side every moment of every day, and if she'd leave the room for a few minutes, he'd glower impatiently until she returned. "Where's Nancy?" he'd shout. "I need her this minute." During one visit he confided to me, "Ben, I've got a great idea for a new spy plane. Get Allen Dulles on the line." I replied,

"Kelly, Dulles passed away many years ago." He became agitated. "The hell you say. You lying bastard. I never did trust you. Get Dulles on the line, I said."

When the Blackbird made its last flight in 1990, I called Nancy and asked her if she thought we could take Kelly out to the plant for the flyby we had scheduled. We both agreed that we didn't want any of his old friends to see him in his condition, but rather to remember him the way he was when they last saw him. So we put him into a limo that had dark windows, making the passengers invisible from the outside, and drove him to the Skunk Works. Kelly was not alert that day, and I really was not sure he understood what that ride from the hospital was about.

All the employees from the Skunk Works were standing outside waiting for the overflight. Kelly sat in the car. We had put out the word that he was not feeling well and would not be able to greet anybody. Everyone respected that and they cheered the car's arrival. Around the time the SR-71 came roaring in over the rooftops and cracked out two massive sonic booms in salute, Nancy had stirred him awake and partially lowered the car window. The booms were as loud as thunderclaps. Kelly looked up, startled. "Kelly, do you know what that was?" I asked. "The pilot was saluting you. We all are saluting you." He didn't reply to my question and seemed to be nodding off. But when I looked at Kelly again he had tears in his eyes.

Kelly Johnson died on the final day before my own retirement as head of the Skunk Works — December 22, 1990. He was eighty years old. We ran a full-page black-bordered ad the following day in the *Los Angeles Times*. It showed the logo of our Skunk Works skunk with a single tear rolling down his cheek.

Clarence "Kelly" Johnson was an authentic American genius. He was the kind of enthusiastic visionary that bulled his way past vast odds to achieve great successes, in much the same way as Edison, Ford, and other immortal tinkerers of the past. When Kelly rolled up his sleeves, he became unstoppable, and the nay-sayers and doubters were simply ignored or bowled over. He declared his intention, then pushed through while his subordinates followed in his wake. He was so powerful that simply by going along on his plans and schemes, the rest of us helped to produce miracles too.

Honest to God, there will never be another like him. He was a great boss if he liked you and a terrible boss if he didn't. Once he was down on an employee, the situation was usually terminal. We would kid, "The only way out of Kelly's doghouse is out the door." Unfortunately, that was true. I was annoyed by things he did at least half a dozen times a week — but I loved that guy.

The fact that there were very few Jews in the top management echelons of aerospace around the time he pushed me to succeed him at the Skunk Works didn't concern him whatsoever. I mentioned the religion question to him in passing. "Ben, I don't give a damn how you pray. I only care about how you build airplanes. And that's all our board will care about too." That was that.

In the mere act of trying to please him and live up to his expectations, I became twice the man I otherwise would have been. Like all the rest of us at the Skunk Works, I ran my heart out just to keep up with him. Kelly, I thank you. All of us do.

15

THE TWO-BILLION-DOLLAR BOMBER

KELLY'S GHOSTLY VOICE nagged at me during the fifteen years I occupied his big corner office and ran his Skunk Works. I always thought of the place as his, because his personality and character were branded on everything we did. Whenever I did something I knew he would never approve of, the old pain in the butt would be hammering at my conscience with a sledgehammer. Expediency and Kelly were archenemies. As his successor, I inherited all his old nemeses as well as his friends. All of Kelly's fourteen golden rules for running the Skunk Works stayed in place: they worked for him and they worked equally as well for me.

Angels belong in heaven, not in the tough competitive world of aerospace, but I kept my word to Kelly and never did build an airplane that I didn't believe in. Like him, I turned down projects I felt were wrongly conceived. I never lied to a customer or tried to dodge the heat when we screwed up. I knew how other companies operated, and I was convinced that our reputation for integrity would gain more business than we would ever lose by turning away questionable ventures. And I was right.

I admit that being extremely careful about how we spent our time and limited resources was easier when profits were high and our workforce fully engaged. But I also turned back bucks when business was lousy too. For example, during one very slack period in the mid-1980s, the Reagan administration was ready to sign up the Skunk Works on a three-year feasibility study for developing a hypersonic airplane, which, by definition, meant an aircraft capable of flying faster than five times the speed of sound. The Reagan science advisers were proposing an airplane that flew at Mach 12 and offered me a million dollars per Mach number to show them how it could be done. The trouble was that I couldn't design such a vehicle if they offered me twelve *billion*. That project was nothing but a simple-minded boondoggle from start to finish.

President Reagan had proposed a national hypersonic plane project during a televised address. The way he described it, Flash Gordon might have been his speechwriter. The vehicle, a commercial passenger plane, would take off from a regular airport runway, climb above the stratosphere into space, then, as an intercontinental rocket vehicle, blast into orbit, before gradually descending like a regular airliner to a distant airport. Reagan called the hypersonic plane "The Orient Express" because it would fly from New York to Tokyo in only two hours. Reagan wanted to build it in four to eight years. He'd be lucky to do it in fifty.

I was outraged by that speech — not at the president, but at his technical team, which apparently had sold him a hypersonic version of the Brooklyn Bridge. I phoned Reagan's chief science adviser at the White House, Jay Keyworth, and told him the idea was utterly absurd. I reminded Keyworth of the enormous problems we had encountered building the Blackbird, which flew "only" at Mach 3.2. I said to him, "Do you know what would've happened if we tried to fly much faster than that? The airplane's surface would have come apart from heat friction. And that was titanium. Do you have something stronger? And by the way, our crew wore space suits and we still worried about boiling them alive if our air-conditioning system failed. And you are proposing to fly at Mach 12, where the surface heat on the fuselage would be 2,500 degrees and still have a passenger cabin filled with women, children, and businessmen, sitting around in their shirtsleeves! Not in my lifetime. Nor in yours." I told Keyworth, "Whoever dreamed up that presidential address ought to be canned. I'm not at all certain we would have that kind of technology ready by the middle of the damned twenty-first century, and if you don't realize that, you are in the wrong business."

But the lure of building a hypersonic airplane dies hard and has become fool's gold to aerospace dreamers. The idea is still kicking around in Congress, its proponents in search of funding like a stray dog sniffing around for a bone. Building the airplane, to be called the X-30, will be a joint project of NASA and the Defense Department. But long before the first serious dollar is plonked down, someone in charge had better realize that Reagan's "Orient Express" is really two separate concepts — one a rocketship and the other an airplane. Most likely, that particular twain shall never meet successfully.

Do the virtuous get their just rewards? The short answer is not if they're dealing with the Pentagon on a regular basis. I had thought, for example, that because the Skunk Works had performed so brilliantly in developing the Stealth F-117A fighter in the mid-1970s that we had earned post position on the inside track for a new stealth bomber. To me, that was a logical evolution from one highly successful program to another. Events not only would prove me wrong, but would lead to the most costly debacle in the history of the defense industry.

I began with the best of intentions: to interest the bluesuiters in a stealth bomber that could carry out a mission over the most heavily defended targets. I made my pitch at the Pentagon one spring day in 1978, during lunch with two of the sharpest people in the business — Gene Fubini, head of the Defense Science Board, and Defense Under Secretary Bill Perry, who was the Carter administration's czar of research and development and godfather of our stealth fighter program. Perry's boss, Defense Secretary Harold Brown, had given him control over stealth. Both Bill and Gene were really depressed over the costly delays and poor testing of Rockwell's new B-1 bomber, built to replace the Strategic Air Command's aging B-52 bomber fleet, but which had the look and feel of a lemon.

The Strategic Air Command was thinking of going instead with an updated version of the General Dynamics F-111, the swing-wing tactical fighter-bomber, which in its development phase as the TFX had been mired in political controversy and cost overruns of horrendous proportions. General Richard Ellis, who now headed SAC, looked favorably on the idea. Since the Air Force bought airplanes by the pound, Ellis wanted a smaller bomber that would cost less and be bought in large numbers in spite of tightening defense budgets. It was far from a perfect solution to SAC's need to update its bomber command, but at that point most blue-suiters were struggling to find an alternative to being stuck with squadrons of B-1s that seemingly had a decade's worth of problems to solve before they could become operational.

"If you guys are eager for a small bomber," I told Fubini and Perry, "look no further than our basic design for the stealth fighter. All we've got to do is make it larger and we have an airplane that could carry the payload of the F-111, but with a radar cross section at least ten orders of magnitude better. We'll hit the most heavily defended target on your list. Can the F-111 make the same claim?"

Both Perry and Fubini knew damned well I wasn't just blowing sales smoke. They were both privy to the extraordinarily low radar results we were achieving with the early models of our fighter. Perry had also recently signed off on a study contract for us to begin designing a stealth naval vessel. He was practical and hard-nosed and demanded results, but he considered us the industry leader in the new stealth technology. Still, Perry had no intention of granting us a monopoly on stealth.

But canceling the B-1 bomber — rendered obsolete by stealth — was a major political mess. It would cost Rockwell millions of dollars and more than ten thousand jobs at its Palmdale, California, plant, and was certain to stir an explosive protest by California's large and powerful congressional delegation. But before our lunch broke up, I had the clear impression that Perry was going to suck up his courage and push Harold Brown and Jimmy Carter to cut their losses and shelve the B-1, which had been designed principally to nuke the Russians by coming in low on the deck to make its bombing run, avoiding radar detection. The Air Force had completed a disturbing study of the airplane's survivability against the latest Soviet ground and air weapons that indicated that 60 percent of the B-1 attack force would be shot down before reaching its target. That loss rate was intolerable. By contrast, the Skunk Works had commissioned an independent study by a defense think tank showing that a bomber employing our stealth technology would achieve a survivability rate over the most heavily defended targets of greater than 80 percent.

A few days after my Washington lunch with Fubini and Perry, I received a call on my secure line from Major General Bill Campbell, who was head of planning at SAC. Bill was no stranger to the Skunk Works. He was a former SR-71 pilot, and I knew him well. "Ben," he began, "General Ellis would be very receptive to a stealth bomber. I want to send out to the Skunk Works a couple of our most senior bomber pilots to sit down with you and your people and work up for General Ellis's approval the requirements for a deep-penetration stealth attack bomber."

I was delighted. SAC's needs and our technology would be in perfect sync from the earliest planning stages. And ever mindful that the final decision rested with General Ellis, I nicknamed our stealth bomber project after Ellis's wife, Peggy, and I hoped he would remain happily wedded at least until we had the contract in hand.

The SAC pilots, both colonels, worked with us in Burbank for two months, helping to draw up a requirement description for a small tactical bomber with a range of 3,600 nautical miles, carrying a payload of 10,000 pounds. Our airplane was configured to supplant its potential rival, the F-111.

General Ellis quickly approved the program. And we received funding for a two-year development study. So I had every reason to believe that the Skunk Works was going to stay busy for many years to come. We would soon be building squadrons of stealth fighters, maybe as many as 150 of them, and almost simultaneously begin producing a similar number of stealth bombers — and I was anticipating more work than even Lockheed's main plant could possibly handle.

The bomber project alone would be enormous. And it never crossed my mind that we might still lose the bomber project after Bill Perry convinced President Carter to kill the B-1. We were the logical choice to replace the B-1.

Perry was convinced that a stealthy deep-penetration bomber would give us air supremacy over the Soviet bloc for at least a decade or longer. He sold Secretary Brown and the Joint Chiefs. They, in turn, sold the president. Anti-stealth technology was a hundred times more difficult to develop than the original stealth technology itself, and would demand extraordinary breakthroughs in the area where the Russians were at their weakest — in supercomputers.

That period at the Skunk Works was the busiest I had ever been. Had I been less preoccupied juggling several big stealth projects simultaneously I might have given more thoughtful consideration to life without Bill Perry at the Pentagon. Because as the presidential campaign heated up and we headed into the fall election, President Carter was clearly in deep political trouble and the chances were growing that Ronald

Reagan was about to become the new commander in chief. Perry was a Democrat and was certain to be replaced by the Reagan defense team. Perry enjoyed respect both in the Pentagon and on the Hill for his technical acuity; without him, the Skunk Works lost a true believer in stealth technology, willing to push against the Pentagon bureaucracy to get important work done.

Northrop was our closest rival in stealth technology. Although they had lost to us in the stealth fighter competition, they were damned good. Their stealth guru was a bearded maverick named John Cashen, a shrewd and tough competitor, who once told me over a few friendly beers that if he had a choice between going to bed with the world's most beautiful woman or beating the Skunk Works out of a contract, he would not hesitate for a second knowing which to choose. "I'd rather screw Ben Rich any time," John chuckled.

John had heard rumors about our supersecret bomber project and managed to push his way into competition with an unsolicited proposal of his own. "This is going to be a huge project, in the billions of dollars, and we can't just hand it to you on a platter," an Air Force general told me. That was probably true, but I knew it was only half the story.

The open secret in our business was that the government practiced a very obvious form of paternalistic socialism to make certain that its principal weapons suppliers stayed solvent and maintained a skilled workforce. Aerospace especially demanded the most trained workers, a labor pool totaling about a quarter million, in the employ of the four or five biggest manufacturers and their host of subcontractors. Each of the major players enjoyed its own special niche, which kept contract awards relatively equitable. The largest was McDonnell Douglas, which specialized in fighters, building hundreds of F-15 interceptors for the Air Force and the Navy's top fighter, the F-18. Next came General Dynamics, builder of the

F-16, a cheap, lightweight fighter sold by the hundreds to our NATO allies, as well as submarines, tanks, and missiles. Lockheed was a solid third, specializing in Polaris missiles, satellites, military cargo aircraft, and spy planes. And finally, Northrop and Rockwell.

At the time that the blue-suiters informed me that Northrop would be competing against us for the stealth bomber, the rumor in the industry had Northrop taking it on the chin with big losses on a project I was familiar with. I couldn't help but chuckle because they had screwed up royally while trying to peddle the lightweight fighter they had wanted me to come aboard to build.

Kelly was right on two counts: Northrop never did start up a Skunk Works operation and its top management was all over that lightweight fighter, interfering in ways that made a bad situation infinitely worse. They had lost more than $100 million on that single-engine fighter, called the F-20, built at the administration's suggestion as a so-called nonprovocative fighter, which meant one that was made to be sold to friendly countries but designed to be vulnerable to our own state-of-the-art interceptors. Arming our friends was good business, but being able to shoot them down if they became our enemies was good strategy. To build this kind of airplane required the permission and cooperation of the administration, which could otherwise block such hardware sales.

So Northrop zeroed in on the Taiwanese, who were receptive to upgrading their fighter squadrons with the new Northrop product. But when the mainland Chinese voiced outrage at the impending sale and called it a serious provocation, the administration got nervous and withdrew Northrop's license to sell the fighter.

Perhaps tacitly acknowledging the administration's culpability in the fighter fiasco, the Pentagon invited Northrop into the bomber competition. I should have read the tea leaves right then about the final outcome of the competition, but I was naive and perhaps a trifle self-confident that we would win on merit, given our expertise and experience in stealth technology. We had the better team, but Northrop had the greater need.

Rockwell was already on the ropes because of the loss of the B-1, and if Northrop was forced to endure big layoffs and big losses with no project in sight to turn things around, the impact would devastate the aerospace business and create economic and political turmoil. Carter and the Democrats were standing in the middle of a thawing lake watching the ice crack all around them.

Because of stealth, we in the Skunk Works had actually prospered under the Democrats. In fact, during his final months in office before the Reagan team took over, Bill Perry was deeply concerned that we in the Skunk Works were overextended with stealth projects and were underestimating the demands involved on our facilities and workforce if we won the bomber competition. About two weeks before Reagan's inauguration, Bill called me to offer guidance. "Ben," he cautioned, "this project will be too big for you to handle in the usual Skunk Works way. The Air Force will want five bombers a month. You'll need ten times the space and workforce. I just don't think you can go it alone. Get yourself a partner and team up."

His suggestion made sense, and I phoned Buzz Hello, who was head of Rockwell's aircraft division, which had been decimated by the administration's decision to cancel the B-1 and had thousands of square feet of empty floor space at its big assembly plant in Palmdale, about sixty miles from Burbank. I enjoyed pleasant relations with Hello, and using his giant facilities would save us a fortune. And he had on payroll a lot of skilled workers who had already been security cleared by working on the B-1. Hello was extremely receptive to my call and agreed to become my partner on the stealth bomber. Five minutes after I hung up, Hello received a call from Northrop's CEO, Tom Jones, asking him to be Northrop's partner on the same project. Jones was furious that I had beaten him to Rockwell and partnered with Boeing instead.

Teaming up was a great way to economize and cut financial risks, but it was also tricky. Today's teammate was tomorrow's competitor and there was a natural reluctance on both sides to share certain state-of-the-art technology or advanced production techniques. I took thirty of Buzz Hello's bomber engineers and brought in Lockheed's best program manager, Dick Heppe, to take charge for us, figuring that after we won the competition, Lockheed's main plant would share a lot of the B-2's production with Rockwell.

After Perry left, it didn't take long for the project to change drastically. He was barely out the door when all the Air Force stealth programs were moved out of the Pentagon to Wright Field in Dayton, under General Al Slay, head of the Air Force Systems Command. Slay was a true believer in the new stealth technology, but he immediately changed the game plan. He was not at all interested in our original requirement for a medium-size bomber; he wanted a big bomber with expanded payload and weapons capability and a six-thousand-mile range. We rushed back to the drawing board to meet his demands.

The funding for the competition came out of a secret stash in the Air Force budget. A young colonel working in the Air Force "black program" office at the Pentagon, named Buz Carpenter, arbitrarily assigned the funding the code name Aurora. Somehow this name leaked out during congressional appropriations hearings, the media picked up the Aurora item in the budget, and the rumor surfaced that it was a top secret project assigned to the Skunk Works — to build America's first hypersonic airplane. That story persists to this day even though Aurora was the code name for the B-2 competition funding. Although I expect few in the media to believe me, there is no code name for the hypersonic plane, because it simply does not exist.

Northrop's design team and mine worked in total ignorance of what the other side was doing. But following basic laws of physics, they came up with strikingly similar designs — a flying-wing shape, a type that Jack Northrop, the company's founder, pioneered in the 1940s. Unfortunately, he was not able to create aerodynamic stability in the age before sophisticated computerized avionics, so his flying wing never really got off the ground. Our designers and Northrop's had no problem making our wings fly and concluded that this unusual boomerang shape afforded the lowest radar return head-on and provided the favorable lift-over-drag ratio necessary for fuel efficiency in long-range flight.

I had absolutely no idea that we and Northrop were designing the same basic airplane shape until Gene Fubini, head of the Defense Science Board, came to visit me at the Skunk Works and saw a model of our stealth bomber on my desk and gasped, "How in hell did you get a model of the Northrop stealth bomber?" When Fubini saw my amazed look, he knew he had spoken indiscreetly.

Both companies were preparing quarter-size models of their designs for a "shoot-out" on an Air Force radar range in New Mexico to determine stealthiness. Wind tunnel tests would follow to determine the best lift-over-drag ratios and other aerodynamic characteristics. The winner would claim the big bomber contract, a high-stakes twenty-year B-52 replacement program.

Although the two designs were very similar, the big difference between them was that John Cashen was getting advice from a three-star general at the Pentagon to make the airplane as large as possible to extend its range, while I was listening to a three-star general at SAC headquarters in

Omaha, who urged me to stay as small as I could get away
with while still meeting the basic Air Force requirements for
the new bomber. "I'm telling you, Ben, that small will win
over big, because budget constraints will force us to go with
the cheaper model in order to buy in quantity." His strategy
made sense. But it also created a few significant structural
differences in the two models. Because our airplane was designed to be smaller, the control surfaces on the wing were
smaller, too, which meant we needed a small tail for added
aerodynamic stability. Northrop had larger control surfaces
and needed no tail at all. So they had a slight advantage in lift-
over-drag ratios, which meant a better fuel efficiency for
extended-range flying.

In May 1981, we and Northrop contested on the Air Force
radar range. Our results were spectacular; through the grape-
vine I heard that we beat John Cashen across the board, on all
frequencies. A few weeks later I received a classified message
from Wright Field questioning the figures we had submitted
on aerodynamic wing efficiency. The message was addressed
to Northrop, but mistakenly routed to me. So I saw that
Northrop's team was claiming an efficiency 10 percent greater
than our own. Frankly, I would question that, too.

Our quoted price to the Air Force per B-2 was $200 million.
I heard that Northrop's quote was significantly higher, so I was
shocked when we received formal notification, in October
1981, that Northrop had been awarded the B-2 project "on the
basis of technical merit." I was so outraged that I took the
unprecedented step of trying to challenge the ruling. Lockheed's CEO, Roy Anderson, agreed with me and marched on
Verne Orr, then secretary of the Air Force, to protest. The two
had an angry confrontation. Orr pounded on his desk and
said, "Goddam it, not only was Northrop better than you, they
were *much better* than you." Anderson, barely in control of his
own temper, just looked Orr in the eyes and responded, "Well,
Mr. Secretary, time will tell."

Yes, indeed. Truer words than Roy Anderson's were seldom spoken.

A blue-suiter called me to explain that the Air Force had determined that Northrop's B-2 had better payload and more range and therefore would be the better buy. While it was true that Northrop's B-2 was more visible in most radar frequencies than our airplane and therefore more vulnerable, it would need to make fewer sorties because it carried more bombs than our model. Therefore, fewer sorties evened out our advantage in being less visible. "The bigger the bomber, the fewer the missions over hostile territory. Their loss rate would be no worse than yours, and might even be better," Secretary Orr had insisted to Roy Anderson. I figured Orr must have had Jesuitical training.

Lockheed's management was, of course, disappointed, but no one blamed me for the loss. To a man, we knew we deserved to win that contract. But the toss of the dice was out of our hands, and we found solace in our leadership in stealth technology, which had made the Skunk Works a billion-dollar enterprise for the first time in its history.

In the end, the government's B-2 decision cost the taxpayers billions. Northrop was supposed to build 132 B-2s at a cost of $480 million each — more than twice what we had originally estimated per airplane. But as those projected costs mounted drastically, Congress lowered the number of bombers to be built to seventy-five and the cost per airplane leaped to $800 million. The fewer the airplanes, the higher the cost is a reliable rule of thumb and a painful lesson about the awful cost of failures in the expensive defense industry business. Now the number of B-2s authorized by Congress is only twenty, and the American taxpayers are spending an incredible $2.2 billion on each B-2 being produced, making the B-2 the most expensive airplane in history. When one crashes — and new airplanes inevitably do go down — it will be not only a tragedy but a fiscal calamity. Northrop's management is in large part to blame for all the delays and cost overruns, but so is the Air Force bureaucracy, which has swarmed over this project from the beginning. When we began testing our stealth fighter, the combined Lockheed and Air Force personnel involved totaled 240 persons. There are more than two thousand Air Force auditors, engineers, and official kibitzers crawling all over that troubled B-2 assembly building in Palmdale. What are they doing? Compiling one million sheets of paper every day — reports and data that no one in the bureaucracy has either the time or the interest to read.

The Air Force now has too many commissioned officers with no real mission to perform, so they stand around production lines with clipboards in hand, second-guessing and interfering every step of the way. The Drug Enforcement Agency has 1,200 enforcement agents out in the field fighting the drug trafficking problem. The DOD employs 27,000 auditors. That kind of discrepancy shows how skewed the impulse for oversight has become both at the Pentagon and in the halls of Congress.

Under the current manufacturing arrangements, Boeing makes the wings, Northrop makes the cockpit, and LTV makes the bomb bays and back end of the B-2 airplane, in addition to four thousand subcontractors working on bits and pieces of everything else. Because of the tremendous costs involved, this is probably a blueprint for how big, expensive airplanes will be built in the future. For better or worse, this piecemeal manufacturing approach — rather than the Skunk Works way — will characterize large aerospace projects from now on. With many fewer projects, the government will have to spread the work across an ever broader horizon. What will happen to efficiency, quality, and decision making? At a time of maximum belt-tightening in aerospace, those are not just words but may well represent the keys to a company's ability to survive.

Other Voices
Zbigniew Brzezinski
(National Security Adviser to President Carter)

When our administration canceled the B-1 bomber program, we knew we would be attacked by political opponents who were unaware of tremendously promising breakthroughs there on stealth technology. Both developments rendered nearly obsolete everything about the B-1, and we in the administration saw that they represented the way of the future, that they were viable, and given what we knew about the state of play in stealth projects and the record of performance by the Skunk Works, that they were really going to perform as advertised. But because of national security, we were unable to reveal to the public the existence of stealth and exploit the strategic facts about it that influenced the decision we made to cancel the B-1. Planning had already begun on a whole new series of stealth bombers and fighters that would revolutionize aerial warfare. So we bit the bullet and just took the heat. This was similar to a political problem faced twenty years earlier by President Eisenhower, who was unable to reveal the U-2 overflights of Russia to answer the charge of a so-called missile gap made against him in the 1960 election campaign.

Leapfrogging technology was the name of the Skunk Works' game, and that occasionally created political problems for both the administration and Congress. That amazing Skunk Works organization was unique in the world in its ability for stretching far beyond that which was thought to be feasible and enjoying a success rate unprecedented for advanced technology projects. Now, in the post–cold war era, we are likely going to be involved in a variety of future conflicts in which overflights for intelligence purposes and for military operations will be of enormous importance. That access is going to have to be surreptitious and undetectable, and clearly the Skunk Works will be continually called upon to keep leapfrogging technology in behalf of the national security. But how we will be able to maintain the tremendously high standards of the Skunk Works during a new era of downsizing defense and intelligence appropriations is really outside my realm of expertise. What is clear is the nation's need to keep this kind of unique operation intact and thriving far into the foreseeable future. Downsizing notwithstanding, it simply must be done.

16

DRAWING THE RIGHT CONCLUSIONS

IN MY FORTY YEARS at Lockheed I worked on twenty-seven different airplanes. Today's young engineer will be lucky to build even *one*. The life cycle of a military airplane is far different from the development and manufacturing of anything else. Obsolescence is guaranteed because outside of a secret, high-priority project environment like the Skunk Works, it usually takes eight to ten years to get an airplane from the drawing board into production and operational. Every combat airplane that flew in Operation Desert Storm in 1991 was at least ten to fifteen years old by the time it actually proved its worth on the battlefield, and we are now entering an era in which there may be a twenty- to thirty-year lapse between generations of military aircraft.

Whether we as a nation will develop intelligent military planning and spending policies in the post–cold war era, I will leave to futurists and politicians to argue about. My interest is the future survival of the Skunk Works as a widely adopted concept, and understanding the key reasons for its unsurpassed success. Will its future be anywhere as bright as its past? Will the concept itself finally receive broad acceptance across the industrial landscape at a time when development dollars are as sparse as raindrops on the Sahara?

To my mind, the leaner and meaner Washington becomes in doling out funding for defense, the more pressing the need for Skunk Works–style operations. Any company whose fortune depends on developing new technologies should have a Skunk Works in operation; in all, there are fifty-five or so scattered around various industries, which isn't very many. But if Lockheed's Skunk Works has been a tremendously successful model, why haven't hundreds of other companies tried to emulate it? One reason, I think, is that most other companies don't really understand the concept or its scope and limitations, while many others are loath to grant the freedom and independence from management control that really are necessary ingredients for running a successful Skunk Works enterprise.

Unfortunately, the trend nowadays is toward more supervision and bureaucracy, not less. General Larry Welch, the former Air Force chief of staff, reminded me recently that it took only two Air Force brass, three Pentagon officials, and four key players on the Hill to get the Blackbird project rolling. "If I wanted an airplane and the secretary of the Air Force agreed," the general observed, "we had four key congressional committee chairmen to deal with and that was that. The same was true of the stealth fighter project — except we had eight people to deal with on the Hill instead of four. But by the time we were dealing with the B-2 project, we had to jump through all the bureaucratic hoops at the Pentagon and on the Hill. So it is harder and harder to have a Skunk Works."

To buck smothering bureaucratic controls inside or outside government takes unusual pluck and courage. Smallness, modest budgets, and limiting objectives to modest numbers of prototypes are not very rewarding goals in an era of huge multinational conglomerates with billion-dollar cash flows. There are very few strong-willed individualists in the top echelons of big business — executives willing or able to decree the start of a new product line by sheer force of personal conviction, or willing to risk investment in unproven technologies. As salaries climb into the realm of eight-figure annual paychecks for CEOs, and company presidents enjoy stock options worth tens of millions, there is simply too much at stake for any executive turtle to stick his neck out of the shell. Very, very few in aerospace or any other industry are concerned about the future beyond the next quarterly stockholders' report.

Yet if times stay tough and the New World Order evolves without any new big-power confrontations, the need for innovative, rapidly developed, and relatively inexpensive systems that are best supplied by a Skunk Works will be greater than ever. Which is the main reason why I remain optimistic about the future viability of the Skunk Works. By its very definition as a low-overhead, advanced development operation for crash production of hot items — prototypes representing cutting-edge technologies that the customer eagerly needs or wants to exploit — the Skunk Works is needed more than ever. The beauty of a prototype is that it can be evaluated and its uses clarified before costly investments for large numbers are made.

Extremely difficult but specific objectives (e.g., a spy plane flying at 85,000 feet with a range of 6,000 miles) and the freedom to take risks — and fail — define the heart of a Skunk Works operation. That means hiring generalists who are more open to nonconventional approaches than narrow specialists. A Skunk Works is allowed to be less profitable than other divisions in a corporation only if its projects are not financial back-breakers and are limited to producing about fifty units or so. Going "skunky" is a very practical way to take modest risks, provided that top management is willing to surrender oversight in exchange for a truly independent operation that can make everyone look good if its technology innovations really catch on, as with stealth. By keeping low overhead and modest investment, a Skunk Works failure is an acceptable research and development risk to top management.

Frankly, in today's austere business climate I don't think a Skunk Works would be feasible if it could not rely on the resources of the parent entity to supply the facilities, tools, and workers for a particular project and then return them to the main plant when the task is completed. But given the right project and motivation, even the usually rigid corporate controls dominating Detroit automakers can benefit enormously from a Skunk Works–style operation for new product research and development, as Ford Motor Company proved recently in producing a new model of its classic Mustang automobile.

For many years the idea of reopening that once popular line of cars was rejected by company executives as being prohibitively expensive. Development costs were projected at more than $1 billion. But in 1990, management put together an ad hoc Skunk Works operation called Team Mustang, composed of designing and marketing executives and expert shop people, swore them to secrecy, then instructed them to design and produce a new Mustang for 1994. Most important, management allowed Team Mustang to do the job with a minimum of second-guessing and management interference. The result: the group took three years and spent $700 million to produce a new vehicle that was extremely well received and became one of Ford's hottest sellers. That represents 25 percent less time and 30 percent less money spent than for any comparable new car program in the company's recent history.

General Motors is now following Ford and has started a separate and secret development group of its own for future projects. Four or five aerospace companies now claim to have a Skunk Works. McDonnell Douglas calls its group Phantom Works, and it apparently emulates what we tried to do at Lockheed. Overseas, the Russians and the French have evolved the most sophisticated Skunk Works operations modeled on Kelly Johnson's original principles. The French aerospace company Dassault-Breguet probably has the best operation in Europe. The concept is beginning to spread, and the ground is certainly fertile because there are only a few thousand days remaining before twenty-first-century technology becomes a reality.

At our own Skunk Works nowadays we are investigating development of vertical landing airplanes and the feasibility of hypersonic airplanes to carry space stations into orbit because rockets are so horribly expensive. We are looking into dirigibles as the ultimate heavy-lift cargo transports, perhaps a better and safer way to ship crude oil around the world than vulnerable tanker ships. We could build dirigibles double-hulled for additional environmental safety and fly them high above violent weather systems. We are researching new uses of composite materials used in our stealth airplane and ship prototypes. These materials never corrode or rust and might be used in new bridges. We are also mulling over better methods to speed up the handling and transportation of large volumes of products.

On the military side, the end of the cold war has brought some welcome new thinking into the military development and procurement areas. Avoiding casualties is now a political imperative for any administration, and the search is on to find new nonlethal disabling weapons that can knock out hardware but leave troops relatively unharmed. There are techniques, for example, to disable a tank without killing its crew by using sprays that crystallize metals or supersticky foams that clog traction. Research is underway to perfect powerful sound generators tuned to frequencies that incapacitate humans without causing any lasting harm. This wall of agonizing sound stops troops in their tracks. So can stunning devices and nonlethal gases being researched, as well as laser rifles firing optical munitions that explode dazzling light brighter than a dozen flashbulbs, causing temporary blindness. Another possibility is an immensely powerful jammer that puts out so much energy that it shuts down all enemy communications or missile defense systems.

But schemers never sleep and there are always counters to every new technology. Currently, the French and Germans are trying to create a missile that can shoot down our stealth fighter. It might well take them twenty years to succeed, but ultimately they will find a way. And then we will find a way to counter their way, and on and on — without an end.

Lockheed's Skunk Works will find useful and productive ways to stay in business for years to come. But a Skunk Works is no panacea for much that ails American industry in general or the defense industry in particular. I worry about our shrinking industrial base and the loss of a highly skilled work-force that has kept America the unchallenged aerospace leader since World War II. By layoffs and attrition we are los-ing skilled toolmakers and welders, machinists and designers, wind tunnel model makers and die makers too. And we are also losing the so-called second tier — the mom-and-pop shops of subcontractors who supplied the nuts and bolts of the industry, from flight controls to landing gears. The old guard is retiring or being let go, while the younger generation of new workers lucky enough to hold aerospace jobs has too little to do to overcome a steep learning curve any time soon. I've recently seen young workers install hydraulic lines di-rectly over electric wires — oblivious to the dangers of a hy-draulic leak that could spark a fire. We are not producing enough airplanes for workers to learn from their mistakes.

During the 1980s an incredible *$2 trillion* was spent on mili-tary acquisitions alone, with new aircraft receiving the largest share of the defense budget, about 43 percent, followed in dis-tant second place by missiles and electronics. The develop-ment costs of fighters have increased by a factor of 100 since the 1950s, and unit procurement costs have risen 11 percent every year since 1963! Small wonder, then, that there were only seven new airplanes introduced in the 1980s, compared to forty-nine in the 1950s. Only three new airplanes have been produced so far in this decade. Correspondingly, the aerospace industry has lost more than a quarter of its workforce since 1987.

New technology cannot be put on a shelf. It must be used. And the desperate need is to try to find ways to drastically reduce costs that would allow new generations of aircraft to evolve within the parameters of extremely modest defense expenditures. That will be the great challenge facing the Pentagon and the defense industry in the years to come.

I would like to share a few cost-saving ideas, but with the following caveat: in some ways I really do think that aerospace has gotten a bum rap from its critics. For example, General Motors spent $3.6 billion giving birth to the Saturn automobile, and it doesn't even go supersonic. We spent $2.6 billion creating the stealth fighter and were able to keep costs down by incorporating the flight controls of the General Dynamics F-16 fighter and using the engine from the McDonnell Douglas F-18. We didn't start from scratch but adapted off-the-shelf avionics developed by others. Avionics is the killer expense, costing about $7,000 a pound in a new airplane. A case in point is our own F-22 Advanced Tactical Fighter, which we designed at the Skunk Works in 1988, to replace the F-15, which has been the primary tactical fighter for the blue-suiters since 1972. In answer to the question, do we really need the F-22, comes another question: do we really want our combat pilots putting their lives on the line in a fighter now more than twenty-two years old?

The F-22 is a performing miracle. It can fly supersonic without afterburners, and using a revolutionary Thrust Vector Control System, can fly at extreme angles of attack while changing directions at high speeds, thereby outperforming any other airplane in the world — all this with the stealth invisibility achieved by the F-117A. To perform its incredible feats, the F-22's avionics is as powerful as seven Cray computers! When we bid on this airplane we did so with the understanding that we would build seven hundred of them. That number would justify the enormous development costs that we shared with the government. Uncle Sam gave us $690 million and we put up a similar amount. The development phase was so expensive that we partnered with General Dynamics and Boeing. Northrop, our competitor, also put up $690 million in partnership with McDonnell Douglas. We won the competition, but all five companies involved in the F-22 contest have lost. We, the winners, will never make back our original investment because in the current budget crunch the government has cut back sharply on the number of F-22s it now plans to purchase. Currently, the Air Force has budgeted for four hundred new F-22s, but that number could decrease even further. The fewer the new airplanes produced, the more expensive the unit cost. The F-22 currently costs around $60 million each — the most expensive fighter ever. Meanwhile we and our partners are carrying huge production overheads in tooling and personnel. The sad truth is that our stockholders would have done better financially if they had invested that $690 million in CDs.

New advanced-technology airplanes are budget breakers. The B-2 bomber, at more than $2 billion a copy, proves that point. But we need stealthy long-range bombers like B-2s, which can fly anywhere in the world in twelve hours and drop a payload of forty conventional bombs. Still, unless we manage to get wildly inflated production costs down, aeronautics could well become an arcane art practiced by one or two manufacturers who somehow manage to survive. (Grumman, the Navy's principal supplier of fighters for nearly sixty years, recently abandoned the sluggish fighter business entirely, while several other big manufacturers are also rumored to be planning to quit increasingly high-risk and unprofitable military aircraft production.)

So an urgent question to pose is how can an airplane as revolutionary and advanced as the F-22 be made for considerably less money? One way, which I have been promoting in

vain for several years, would be to make it even more revolutionary than it already is. That is, much of the horrendously
expensive avionics on board allows the F-22 to perform incredible aerodynamic maneuvers, from climbing straight up
to pulling enormous g-loaded turns. But what if we invent onboard weapons systems that perform all the maneuvering
while the airplane itself flies straight and level? In other
words, missiles that could electronically lock onto a dogfighting target and swivel and turn at twelve gs while the airplane
flies at only two gs. Such a reversal would drastically reduce
the huge avionics costs.

Like true Skunk Workers, the aerospace industry as a whole
must start thinking in new directions. Why build each new
airplane with the care and precision of a Rolls-Royce? In the
early 1970s, Kelly Johnson and I had dinner in Los Angeles
with the great Soviet aerodynamicist Alexander Tupolev, designer of their backfire Bear bomber. "You Americans build
airplanes like a Rolex watch," he told us. "Knock it off the
night table and it stops ticking. We build airplanes like a
cheap alarm clock. But knock it off the table and still it wakes
you up." He was absolutely right. The Soviets, he explained,
built brute-force machines that could withstand awful
weather and primitive landing fields. Everything was
ruthlessly sacrificed to cut costs, including pilot safety.

We don't need to be ruthless to save costs, but why build the
luxury model when the Chevy would do just as well? Build it
right the first time, but don't build it to last forever. Why must
every aircraft be constructed to fly for twenty thousand hours
and survive the stresses and strains of a thousand landings
and takeoffs? Why not lower the endurance requirements for
the majority of airframes? Wars are now planned to last
ninety days because after that time ammunition reserves run
out. In battle, most airplanes will be deployed for a few hundred hours at most. It would be cheaper to dispose of them

once they've seen combat than to stockpile vast quantities of replacement parts and engines. We could make a small number of aircraft to last for years in training flights. But produce the majority — the ones destined for a relatively short spurt of combat flying — with less expensive materials. This same sort of Skunk Works' cost-reduction thinking could extend to airplane tires and other parts. Why, for example, must tires last for one thousand landings? If we mass-produced them at somewhat lower standards, we could throw away airplane tires after ten landings and still save money.

We cannot enjoy total product perfection and really don't need it. The only areas where the final result must be 100 percent are safety, quality, and security. That final 10 percent striving toward maximum perfection costs 40 percent of the total expenditure on most projects.

General Electric's jet engine plant at Evendale, Ohio, sells its engines to the commercial airlines for 20 percent less than to the Air Force. Price gouging? No. But the Air Force insists on having three hundred inspectors working the production line for its engines. The commercial airlines have no outside inspectors slowing down production and escalating costs. Instead, the airline industry relies entirely on GE's engine warranty, a guarantee that the engine will function properly or GE will be required to pay a penalty as well as all costs for replacement, repairs, and time lost. Why can't the Air Force operate with similar guarantees and save 20 to 30 percent on engine costs and eliminate three hundred unnecessary jobs to boot?

One of the biggest cost items in defense is logistics management and maintenance. We should reevaluate the design of many components and make them throwaway or limited-shelf-life items. Batteries, brakes, servos, modular avionics, should all be replaced on a definite schedule, not wait for them to wear down. This would reduce the accumulation of large spare part inventories in city-size warehouses, cut down repairs and maintenance, and lower supply pipeline costs. Savings could run in the hundreds of millions.

One area in particular where the Skunk Works serves as a paragon for doing things right is aircraft maintenance. We have proven time and again that the Air Force would be much more efficient using civilian contractor maintenance on its air fleet whenever possible. Fifteen years ago, there were so many mechanical breakdowns on the flight lines at air bases around the world that it took three airplanes to keep just one flying. The reason: lack of good maintenance by inexperienced flight crews. We in the Skunk Works are the best in the business at providing our own ground crews to service and repair our own aircraft. For instance, two Air Force SR-71 Blackbirds based in England throughout the 1970s used Skunk Works maintenance. We had on hand a thirty-five-man crew. By contrast, two Air Force Blackbirds based at Kadena on Okinawa relied on only blue-suiter ground crews, which totaled *six hundred* personnel. Contractors can cross-train and keep personnel on site for years, whereas the military rotates people every three years, and valuable experience is lost.

Currently, two U-2s are stationed in Cyprus with twelve Lockheed maintenance persons, while two other U-2s stationed at Taif, in Saudi Arabia, in support of the UN mission in Iraq, have more than two hundred Air Force personnel. And when the U-2s at Taif need periodic overhauls, they are flown to Cyprus, where our crews do the job.

Another relatively easy cost-reduction scheme would be to rethink aircraft design so that all parts are "no-handed." That is, there would be no left and right hinges or wing flaps or other control surfaces. The cockpit controls would likewise be no-handed. Production learning curves in manufacturing these items would be twice improved by not having to devote half to left and half to right and would reduce significantly spares and storage parts requirements.

Even at the Skunk Works I've seen my share of money wasted, at times in the most ridiculous ways imaginable. One that particularly sticks in my craw occurred when President Johnson first announced in 1964 the existence of the RS-71, the Air Force two-seater Blackbird. That's right, RS-71 was its official designation, but Johnson accidentally turned it around and called it the "SR-71." Instead of putting out a brief correction, the Air Force decided not to call attention to a very minor mistake by the commander in chief and ordered us to change about twenty-nine thousand blueprints and drawings at a cost of thousands of dollars so that they would read "SR-71" and not "RS-71." Another frustrating example was the stubborn insistence of the Air Force to have its insignia painted on the wings and fuselage of the SR-71 Blackbird, even though no one would ever see it at eighty-five thousand feet; finding a way to keep the enamel from burning off under the enormous surface temperatures and maintain its true red, white, and blue colors took our chief chemist, Mel George, weeks of experimentation and cost the government thousands of unnecessary dollars. After we succeeded, the Air Force decided that the white on the emblem against the all-black fuselage was too easy to spot from the ground, so we repainted it pink. Air Force regulations also forced us to certify that the Blackbird could pass the Arizona road-dust test! Years earlier low-flying fighters training over Arizona's desert wastes suffered engine damage from sand and grit. We had to demonstrate that our engine was specially coated to escape grit damage — this for an airplane that would overfly Arizona at sixteen miles high.

Such bureaucratic madness, I am certain, will never entirely disappear no matter how tight things get, but affordability nowadays is even more important than technology, and a genuine attempt to control costs is the highest priority among all branches of the military. The Skunk Works concept as a vibrant force in the American defense industry can come into its own only if and when the government reverses some of its counterproductive practices. And the most obvious place to start in achieving greater efficiency is to ferociously attack unnecessary bureaucratic red tape and paperwork.

I was in Boston recently and visited *Old Ironsides* at its berth, coincidentally at a time when the ship was being painted. I chatted with one of the supervisors and asked him about the length of the government specifications for this particular job. He said it numbered two hundred pages and laughed in embarrassment when I told him to take a look at the glass display case showing the original specification to build the ship in 1776, which was all of three pages.

Everyone in the defense industry knows that bureaucratic regulations, controls, and paperwork are at critical mass and, if unchecked, in danger of destroying the entire system. An Air Force general in procurement at the Pentagon once confided to me that his office handled thirty-three million pieces of paper every month — over one million per day. He admitted that there was no way his large office staff could begin to handle that kind of paper volume, much less read it. General Dynamics is forced by regulations to store ninety-two thousand boxes of data for their F-16 fighter program alone. They pay rent on a fifty-thousand-square-foot warehouse, pay the salaries of employees to maintain, guard, and store these unread and useless boxes, and send the bill to the Air Force and you and me. That is just one fighter project. There are many other useless warehouses just like it. There is so much unnecessary red tape that by one estimate only 45 percent of a procurement budget actually is spent producing the hardware.

Oversight is vitally important, but we are being managed to death and constantly putting more funds and resources into the big end of the funnel to get an ever smaller trickle of useful output from the small end. Over the years in the Skunk Works, we supplied necessary paperwork when it was critically important and eliminated all the rest of it. A Skunk

Works purchase order for vendor development of a system used in an advanced airplane took three pages. The vendor replied with a four-page letter proposal that included specifications for the system under development. And that was that. But at Lockheed's main plant, or at any other manufacturer's, that same transaction typically produced a 185-page purchase order, which led to a 1,200-page proposal, as well as three volumes on technical factors, costs, and management of the proposed project.

To put the paperwork blitz in perspective: there are currently operative throughout the Defense Department, acquisition regulations that reportedly could fill an entire shelf of 300-page books, in addition to 50,000 individual specifications, 12,000 contract clauses for specific components, 1,200 department directives, and 500 separate procurement regulations.

Paperwork should be limited to what the government most needs to keep tabs on. And I cannot deny that over the years the defense industry has had more than its share of cost overruns, bribery scandals, and other serious transgressions, which proves the need for intense scrutiny. In many ways, though, our sullied reputation was somewhat unearned because cost overruns in our industry were seldom more than 20 percent, while in other industries operating entirely by private financing, big-project overruns of 30 percent or even higher are not uncommon.

Nevertheless, excessive government regulation is the penalty we now pay for years of overpromising and lax management in aerospace. At the heart of the defense industry problem was a recognition that if we bid unrealistically low to get a project, the government would willingly make up the difference down the line by supplying additional funding to meet increasing production costs. And it would do so without penalties.

Now if there are serious cost overruns, whether caused by

unexpected inflationary spirals or even "no fault" acts of God, the company is liable to pay for it all or fix any mistakes from its own pocket. The era of the fixed-price contract rules supreme. As for major cost overruns, it is impossible to really surprise the government by suddenly revealing out-of-control expenses because every production line is swarming with government bean counters and inspectors keeping close tabs every step of the way.

Back in 1958, we in the Skunk Works built the first Jetstar, a two-engine corporate jet that flew at .7 Mach and forty thousand feet. We did the job in eight months using fifty-five engineers. In the late 1960s the Navy came to us to design and build a carrier-based sub-hunter, the S-3, which would fly also at .7 Mach and forty thousand feet. Same flight requirements as the Jetstar, but this project took us twenty-seven months to complete. One hint as to the reasons why: at the mock-up conference for the Jetstar — which is where the final full-scale model made of wood gets its last once-over before production — we had six people on hand. For the S-3 mock-up the Navy sent three hundred people. S-3 may have been a more complex airplane than Jetstar, but not thirty times so. But we were forced to do things the Navy Way.

In more recent years the government seemed determined to lower procurement costs through rigorous competition. One curious idea developed by Ronald Reagan's secretary of the Navy, John Lehman, and adapted by the Air Force and other service branches as well, was something called leader-follower competition. The rule of this game was that the winner of a competitive bidding competition had to turn over his winning design to the loser. The loser would then learn how to build the winner's product, and by the third year of production the loser would be allowed to bid on the project against the winner. For example, several years ago Hughes won the competition for an advanced medium-range attack missile against Raytheon, builder of the Patriot missile. The government bought four thousand of these missiles annually. The first year Hughes got the entire order. The second year, Raytheon, which had studied Hughes's winning design, got an order to build one thousand while Hughes dropped to three thousand. By the third year, the government opened up the bidding to full competition and Raytheon won a majority of the buy. They were able to put in a lower bid because they had no initial research and development costs to factor into the price and they took over 60 percent of that missile's production.

The leader-follower concept was an absolute outrage and a debacle. Fortunately, the Pentagon has since abandoned the idea, after admitting that there had to be fairer ways to lower costs, stimulate competition, and spread around new business among winners and losers.

In the so-called New World Order, defense-related procurements will undoubtedly continue to sharply decline into the foreseeable future, rendering the mountains of regulations and the battalions of inspectors and auditors irrelevant. Obviously, we will need continual defense spending and new technologies for as long as the world remains dangerously unstable. But now more than ever, I believe, the Pentagon and industry need to adapt the kinds of specialized managment practices that have evolved out of the Skunk Works experience over the past half century.

My years inside the Skunk Works, for example, convinced me of the tremendous value of building prototypes. I am a true believer. The beauty of a prototype is that it can be evaluated and its uses clarified before costly investments for large numbers are made. Prior to purchasing a fleet of new billion-dollar bombers, the Air Force can intensively audition four or five, learn how to use them most effectively on different kinds of missions and how to maximize new technologies on board. They can also discover how to best combine the new bombers with others in the inventory to achieve maximum combat effectiveness.

One of the problems with our stealth fighter was that because it was hidden for years behind a wall of tight security, most Air Force tactical planners didn't even know it existed and thus had no way of integrating the airplane into overall combat planning and strategy. By the time the F-117A arrived on station for duty in Operation Desert Storm, the airplane was largely a cipher to the high command in terms of its performance capabilities. After the war ended, Lt. General Charles Horner, in charge of Air Force operations, stated frankly that before the F-117A's first combat mission, he was apprehensive about the effectiveness of stealth in combat. "We had a lot of technical data, but I had no way of knowing that we would not lose the entire fleet that first night of the war. We were betting everything on the data proving the technology — but we had no real experience with the airplane to know *for certain* how well it performed under fire. We sent those boys in naked and all alone. As it turned out, the data was right on the mark. But we should've known that before the first attack."

As another streamlining improvement in the years ahead, the government should adopt the Skunk Works' proven procedures for concurrency in manufacturing new airplanes or weapons systems. That is, new weapons systems or airplanes need not be endlessly perfected before production begins, provided that development proceeds carefully, avoiding the messes that both the B-1 and B-2 bombers got into when it was discovered that their avionics and weapons systems, independently produced, just didn't fit into the strategy and design concept of the new bombers. Fixing it cost a fortune. The bottom line in concurrency development is cost savings, provided it is done right. Our experience on the stealth fighter proves it can be cost effective to build in improvements from production model to production model and keep within the budget and time frame contracted for. By the time we built stealth fighter number ten, we had enhanced many features that we were able to quickly install into the first nine models, because we had planned for concurrency from the beginning by keeping detailed parts records on all the production models and designing easy access to all onboard avionics and flight control systems.

Procurement should be on a fast track basis with a minimum of meetings, supervision, reviews, and reports. Whenever possible all parts on a new airplane should be commercially available, not specially made for military specs that are most often overkill and unnecessarily costly.

Another sound management practice that is gospel at the Skunk Works is to stick with reliable suppliers. Japanese auto manufacturers discovered long ago that periodically switching suppliers and selecting new ones on the basis of lowest bidders proved a costly blunder. New suppliers frequently underbid just to gain a foothold in an industry, then meet their expenses by providing inferior parts and quality that can seriously impair overall performance standards. And even if a new supplier does produce quality parts according to the specifications, his parts will not necessarily match those furnished by the previous supplier: his tooling and calibrations might be different, causing the major manufacturer extra costs to rework other system components.

For these reasons Japanese manufacturers usually form lasting relationships with proven suppliers, and we at the Skunk Works do the same. We believe that trouble-free relationships with old suppliers will ultimately keep the price of our products lower than if we were to periodically put their contracts up for the lowest bid.

Still another area of potential cost reduction is security. A classified program increases a manufacturer's costs up to 25 percent. I believe in maintaining tight perimeter security to guard the plant site and keep sensitive materials under lock and key. If we don't allow our people to take out sensitive papers, then we don't have to worry about it. We need to safeguard technologies and weapons systems, but we don't need to hide behind secrecy as a means to cover up mistakes or to block oversight by outside agencies. In the past, the government has slapped on way too many security restrictions in my view. Once a program is classified secret it takes an act of God to declassify it. We should limit its use and be tough about periodic declassification reviews. What was secret in 1964 often is probably not even worth knowing about in 1994. I would strongly advocate reviews every two years of existing so-called black programs either to declassify them or eliminate them entirely. We could save millions in the process.

Just one very practical problem about classification that any reader would immediately understand: how do you transport from Burbank to Washington highly sensitive blueprints or performance studies stamped top secret? Do you call in Federal Express? Do you send them by registered mail? No. You use a special courier, who carries the material in a locked case handcuffed to his wrist. If the material is extremely sensitive, the means of transportation is usually government-chartered flight or use of multiple couriers. So, secrecy classifications are not inconsequential but a burden to all and horrendously expensive and time-consuming. If necessarily in the national interest, these expenses and inconveniences are worthwhile. But we ought to make damned sure that the secrecy stamp is absolutely appropriate before sealing up an operation inside the security cocoon.

Sunset laws on security are an important first step toward real dollar saving. But government has a long list of needed reforms in the area of imposing sunset provisions on dozens of unnecessary regulations. Companies with solid track records should be rewarded with less supervision and outside interference, while companies that fail to meet performance requirements should be penalized severely.

Under existing laws if a company actually brings in a project at considerably less cost than called for in the original contract, it faces formidable fines and penalties for overbidding the project. Not much motivation to save time and money, is there?

All of us in the defense industry would benefit from multi-year funding. The laws require that all procurement allocations be on a strict annual basis, so that when the Air Force reopened the production line on the U-2 back in the early 1970s, for example, and I knew that I would be building about thirty-five U-2s — five a year for the next seven years — I still was prohibited from tooling up for five-years' worth of parts or materials. Not only might the blue-suiters cancel the program, forcing me to eat all those parts for breakfast, but it is against the law for a manufacturer to spend money that has not actually been appropriated. I could be severely fined or penalized by losing the contract entirely if I tried to stock up for several years' worth of materials and parts and tools. But by doing so, I could probably reduce my production costs if I were allowed to purchase in volume tooling and materials for, say, three years at a time, rather than doing so annually and incrementally.

Frankly, I think the government prefers this annual funding system because it can then promise a company the moon and the stars in order to get it to put up significant development capital, then later sharply reduce the procurement, as in the case of the F-22.

One of the bitter lessons of failure that Rockwell learned while building the ill-fated B-1 bomber was that in retrospect they surrendered too much authority and responsibility to the customer. The Air Force allowed Rockwell to build the airframe of the bomber, but the critical onboard avionics, both offensive and defensive, were in the hands of blue-suiters and were uncoordinated and out of sync with the airframe builders. Ultimately, it would cost millions to undo their mistakes. The lesson is that there is no substitute for astute managerial skills on any project. In the absence of effective managers, complex projects unravel. And, by the way, there is an even greater shortage of skilled managers than of effective leaders in both government and industry nowadays. Leaders are natural born; managers must be trained. When Noah designed an ark and gathered his family and a pair of male and female animals of all species to avoid the Great Flood, he demonstrated his leadership. But when he turned to his wife and said, "Make certain that the elephants don't see what the rabbits are doing," he was being a farsighted, practical manager.

I will leave it to historians to debate the effects of our own astronomical defense spending, particularly during the Reagan administration, on the demise of the Soviet Union. Did we really spend them into self-destruction, or did their own corruption and an almost nonexistent functioning economy do the job irrespective of military outlays? Or is the truth somewhere in between? There is no doubt that we ran up enormous debts ourselves that have almost wrecked our own free enterprise system by chasing after enormously costly technologies that were simply beyond our creative grasp. The Reagan administration's Strategic Defense Initiative is a case in point. This was the so-called Star Wars concept of employing an impenetrable defensive shield capable of destroying all incoming enemy missiles launched against us. In an actual all-out nuclear attack, hundreds of missiles would be raining down on us, including many decoys. SDI would instantly acquire them all, distinguish between real and decoys, and save our bacon by knocking out all those nuclear warheads at heights and ranges sufficiently far removed from our own real estate. Some of us thought that SDI stood for Snare and Deception with Imagination, even though it was supposedly the brainchild of Dr. Edward Teller, the father of the hydrogen bomb, who sold the concept to President Reagan. The trouble was that the technologies to make this system function as advertised were not even on the horizon and would ultimately cost trillions of dollars to develop over many, many years. The administration allocated about $5 billion annually in R & D, which did result in some advances in laser and missile research.

Its apologists justify these costs with the claim that the Russians really believed in the seriousness of the administration's intent and were panicked into disastrous spending to try to overcome SDI. If so, we spent a hell of a lot of money in deception and very little in behalf of worthwhile technology. And that will always be the result unless the new technology we are attempting to create is really within practical reach of our current abilities and achievable with reasonable expenditures.

Lockheed's Skunk Works has been a dazzling example of American aerospace at its best, setting the standards for the entire U.S. industry in developing aircraft decades ahead of any others in performance and capability. Fifty years is a long time for a very small development company to stay in business, much less to maintain its unusual competence and morale. But if we are to survive a future every bit as uncertain and turbulent as the past, we will need many more skilled risk-takers like the ones first brought together on the initial Skunk Works project during World War II, to build the first U.S. jet fighter.

We became the most successful advanced projects company in the world by hiring talented people, paying them top dollar, and motivating them into believing that they could produce a Mach 3 airplane like the Blackbird a generation or two ahead of anybody else. Our design engineers had the keen experience to conceive the whole airplane in their mind's-eye, doing the trade-offs in their heads between aerodynamic needs and weapons requirements. We created a practical and open work environment for engineers and shop workers, forcing the guys behind the drawing boards onto the shop floor to see how their ideas were being translated into actual parts and to make any necessary changes on the spot. We made every shop worker who designed or handled a part responsible for quality control.

Any worker — not just a supervisor or a manager — could send back a part that didn't meet his or her standards. That way we reduced rework and scrap waste.

We encouraged our people to work imaginatively, to improvise and try unconventional approaches to problem solving, and then got out of their way. By applying the most common-sense methods to develop new technologies, we saved tremendous amounts of time and money, while operating in an atmosphere of trust and cooperation both with our government customers and between our white-collar and blue-collar employees. In the end, Lockheed's Skunk Works demonstrated the awesome capabilities of American inventiveness when free to operate under near ideal working conditions. That may be our most enduring legacy as well as our source of lasting pride.

A successful Skunk Works will always demand a strong leader and a work environment dominated by highly motivated employees. Given those two key ingredients, the Skunk Works will endure and remain unrivaled for advancing future technology. Of that, I am certain, even in the face of dramatic downsizing of our military-industrial complex. Prudence demands that the country retain a national capability to design and develop both new technologies and weapons systems to meet threats as they arise, especially better and improved surveillance of unstable, hostile regimes.

Nuclear proliferation is a growing menace: a bomb in the hands of the North Koreans, the Pakistanis, or the Iranians makes the world infinitely dangerous and demands the closest surveillance, which only the most advanced technology can provide. At last count there were about 110 local wars or potential trouble spots around the globe to keep a close eye on. Human conflicts may be smaller in size and scope in the post–cold war era, but certainly not in nastiness. Littoral confrontations — local conflicts caused by political, religious, or ideological differences — will probably monopolize international tensions and concerns for at least the remainder of this century. Increasingly, we will be facing small hostile countries armed to the teeth with the latest weapons technologies purchased from irresponsible outlets in Western Europe or from Russian, Chinese, or North Korean sources. A small country firing high-tech weaponry can do as much damage on the battlefield as a major power. Just remind the Russian high command of the tremendous losses of Hind helicopters they sustained in Afghanistan to a bunch of ragtag peasants firing shoulder-held Stinger missiles, supplied to them by our CIA.

Small localized conflicts are going to be played out on the ground by highly mobile strike forces requiring air superiority, overhead surveillance, and surgical air strikes with high-precision guided ordinance; and a Skunk Works that is expert at low rate production of startling new technologies will undoubtedly serve important national security purposes in the future as it has in the past.

Given current contractions in defense spending and needs, the mountainous inventory of big-strike weapons like intercontinental missiles will be sharply reduced and much of it scrapped. The remaining systems will need updating with the latest technologies, improved to reduce maintenance and manpower utilizations — perfect tasks for Skunk Works operations.

As regions of the world become increasingly unstable, the U-2 fleet might undergo its third reconfiguration in its five decades of service to the nation as our preeminent spy plane. And while that is happening a future successor of mine at the Skunk Works will undoubtedly be peddling ideas for solving technological problems arising out of nonuse of weapons — for example, how to keep silo-based missiles reliable and effective after years of sitting inert in the ground. In some cases reliability has dropped below 50 percent. Another big problem that a Skunk Works would be eager to try to solve is eliminating battlefield deaths caused by accidental friendly fire.

Twenty-six percent of our battlefield deaths in Desert Storm
resulted from our own shells and bullets. What is needed is
some sort of foolproof technology, which the Pentagon has
designated IFF — Identify Friend or Foe. The Army plans to
spend nearly $100 million developing exclusive radio fre-
quency signals that troops can use in the field at night (our
GIs may give off a definite buzz), as well as infrared devices
and paints on trucks and tanks that only our side can see
using special lenses.

As the only remaining superpower, the United States will be
wise to resist being drawn into any military intervention on
the short end of public support or lacking a clear threat to our
own national security. But even a leader able to whip up sentiment for "sending in the Marines" will find it dicey to under-
take any prolonged struggle leading to significant casualties.
New technologies will focus increasingly on developing non-
manned fighting machines by using reliable drones, robotics,
and self-propelled vehicles. As we proved in Desert Storm, the
technology now exists to preprogram computerized combat
missions with tremendous accuracy so that our stealth
fighters could fly by computer program precisely to their targets over Iraq. A stealthy drone is clearly the next step, and I
anticipate that we are heading toward a future where combat
aircraft will be pilotless drones. On the ground and at sea as
well, remote-controlled tanks and missile launchers and unmanned computer-programmed submarines and missile frig-
ates will provide the military advantage to those possessing
the most imaginative and reliable electronics and avionics.
Field commanders can conduct battles and actually aim and
fire weapons systems from the safety of control centers thousands of miles away — their targets sighted by the high-
powered lenses aboard drone surveillance aircraft which they
remotely control. This is not a Buck Rogers scenario; this is
around the corner. Tomorrow's most prized military break-

through may be in the form of a dazzlingly new and powerful microchip.

The Skunk Works has always been perched at the cutting edge. More than half a dozen times over the past fifty years of cold war we have managed to create breakthroughs in military aircraft or weapons systems that tipped the strategic balance of power for a decade or longer, because our adversaries could not duplicate or counter what we had created. That must continue to be our role into the next century, if we are to preserve what we have accomplished and be prepared for the hazards as well as the opportunities for the uncharted, risky future.

EPILOGUE
THE VIEW FROM THE TOP

Harold Brown
(Secretary of Defense from 1977 to 1981)

I have thirty years' worth of memories involving the Skunk Works, beginning in the late 1950s while I was at Livermore Laboratories helping out with interpretations of the Soviet nuclear program, which I based in significant part on information available to me that had been obtained by those historic U-2 overflights of Russia. I remember with much less relish May 1, 1960, and the following difficult days during which the story of the U-2 — or at least a part of it — became public knowledge because of the shooting down of Francis Gary Powers. I remember coming to Washington early in 1961 as Director of Defense Research and Engineering and being briefed on the intelligence bonanza achieved by the earlier U-2 missions over Russia. I was amazed by the quantity and its strategic importance.

I was at the Pentagon when the U-2 again played a decisive role, this time during the Cuban missile crisis, obtaining photographs detailed enough to convince first the U.S. decision makers and then our allies of what the Soviets were up to in Cuba. Even though the credibility of our government was much higher in those days than it subsequently has become, I think it would have been extremely difficult for President

Kennedy to carry the public and the allies along with him on the blockade and all its grave implications for nuclear war in the absence of the proof supplied by the U-2 photos.

It was during the mid-1960s that the Blackbird, the SR-71, as the successor to the U-2 began bringing us intelligence information both photographic and electronic. During my watch as Secretary of Defense, we relied on overflights of Cuba by Blackbirds to obtain vital national security information involving the Soviets. These flights enabled us to determine that the Russians were indeed shipping in their latest attack fighters in violation of their 1962 pledge not to deploy offensive weapons in Cuba. Armed with photographic proof, we obtained Soviet assurances that these aircraft were for defensive purposes only. Another overflight in 1979 revealed Soviet troops and armor deployed in Cuba. Caught by our cameras, the Russians claimed they were merely staging maneuvers with their Cuban allies. To counter this, we staged our own training exercise in the vicinity that dwarfed theirs in both size and strength. So in its ability to be dispatched quickly anywhere in the world, the Blackbird proved to be invaluable.

Finally there is the F-117 stealth fighter, which began in development stages during my last years in office. It was a remarkable achievement and excited the imagination of operational planners who finally had the good sense to come up with a workable doctrine and operational concept, combining the airplane's invulnerability with high-precision bombs. The airplane's success in Desert Storm was unprecedented and overwhelming, justifying all our early enthusiasm about the new technology and my decision to pin a medal on Ben Rich's chest during my last days in office.

Caspar Weinberger
(Secretary of Defense from 1981 to 1986)

Airplanes produced by the Skunk Works made enormous contributions to our winning the cold war, and the whole idea of putting money into such a splendid R & D operation for extremely advanced aircraft must continue to be a vital part of our national defense into the next century. So I have very strong views about the Skunk Works. In fact, one of the concerns I have about the current administration is the cuts in R & D funding, because our great strength is the variety and degree of sophisticated weapons that enabled us to do the kinds of things we achieved in the Gulf War. You don't get those kinds of systems the morning after the war starts: you have to spend seven years in development, as we did with the stealth fighter, which performed so flawlessly and with such devastating effect.

The Soviets had great trepidations about that stealth fighter. It was one of three key items on their counter-strategy agenda: the first was NATO and their efforts to decouple us from Europe and see us pull out; the second was to do all in their power to stop work on the Star Wars defense initiative, and the third was to uncover our secrets about stealth. They knew we were working on it because Mr. Carter announced it a few months before the 1980 election in order to pick up some votes. From there on in, their espionage and spy satellite activities directed against the Skunk Works increased tenfold. They were panicked because there was no way they could counter this technology. Just like the Blackbird, that was so far ahead of its time that to this day it could still outperform anything else around, the F-117 was an attack system the other side simply could not find a way to stop. It was designed for surgical strikes against heavily defended targets — get in and out very quickly and not be vulnerable.

I had a stubborn determination to keep that airplane under wraps for as long as possible. I didn't want to use it prematurely, as in our strike against Kaddafi, because I didn't want to risk having a plane go down and having its technology explored. Our military advantage has always been the variety and degree of the sophisticated weapons systems we have been able to produce — systems far ahead of the other side's capabilities. On this score, the Skunk Works is without peer in the world — the airplanes they produced were decades ahead of then current capabilities. To this day, no airplane can begin to outperform their incredible Blackbird. And as for the F-117A stealth fighter, not one of them suffered a single bullethole despite undertaking hundreds of tough combat missions in the Gulf War. This was a very intense conflict, and I was frankly surprised by its tremendous performance, because it was not precisely the kind of mission that the stealth fighter had been designed for.

The stealth program was a classic example of an R & D triumph of historic proportions. We could use the stealth fighter to take out the other side's most effective and dangerous missile sites in a series of quick surgical attacks, leaving them defenseless against our main force of attack aircraft. Stealth changed all the dimensions and equations for planning future air campaigns.

The Skunk Works has been extraordinarily impressive over the years and absolutely vital to the national interest. Every administration since Eisenhower's has counted on them to help keep us in the technological forefront in a dangerous world. So it is against this background that I view with alarm the problem of downsizing our defense industry to the point where the Skunk Works will lose valuable workers and be allowed to dangerously atrophy. The cold war may be over, but the world is certainly not benign. More than ever, it seems to me, we need to continue developing new generations of surveillance aircraft to replace the Blackbirds and U-2s — airplanes that are the hallmarks of the Skunk Works' genius — to attempt to deal with regional conflicts, or to find out whether anxiety-causing military maneuvers in some trouble spot are not actually a prelude to a full-scale military engagement, or just to keep a sharp eye on what North Korea is really up to. And we can't do this vital job effectively just with overhead satellites. We need spy planes — whether manned or ground-controlled drones.

But I am frankly concerned that Lockheed and other companies are looking to new civilian areas because there is no future in defense. General Dynamics is already out of the military aircraft business, and we might be down to two or three main contractors. That means we will have an incredibly hard time trying to hold on to our best and most skilled aerospace engineers and shop workers. Clearly we need a president with the political courage to continue fielding a strong defense and fund it, instead of drastically cutting back and then using the savings for politically popular programs. The Skunk Works has kept us preeminent, and we have got to keep them preeminent.

William J. Perry
(Secretary of Defense for President Clinton)

The Skunk Works enjoys a preeminence in aerospace development based largely on unbounded enthusiasm and a willingness to tackle just about anything. That was amply demonstrated to me back in the 1970s, during my first tour of duty at the Pentagon as undersecretary, while working with Ben and the Skunk Works to develop the stealth fighter. Those guys showed the most extraordinary confidence in their willingness to take on projects that most other companies would avoid because the chances to succeed were too small or too difficult. But the Skunk Works applied their formidable technical skills in very practical ways that no other company was able to match. That, plus their tremendous energy and willingness to work around the clock until a problem was solved, enabled them to enjoy a very close and trusting relationship with the customer that was unique between government and industry.

Without question, my work with the Skunk Works on developing the stealth fighter is the most significant project I have ever been involved in. Back in the seventies, I never would have been able to predict in detail about where we would be in stealth developments in the 1990s. But I did have a vision that what the Skunk Works had produced was the most important military aerospace technology since the invention of radar in World War II. And like radar, stealth would change the way that all subsequent air wars would be fought. I believed then that it was only a matter of time before every new airplane and missile that we built would incorporate stealth features and that stealth would provide our Air Force with a highly leveraged capability to wage air wars. All of that was very clear to me twenty-two years ago, and I can remember the precise moment when I became certain about the real significance of our stealth work — a specific summer day in 1977, when the Have Blue stealth prototype built by the Skunk Works first flew over a radar-measurement test range. The results were just stunning. Have Blue proved that the incredible low radar cross section previously achieved by a wooden model sitting on top of a pole could be duplicated by a combat aircraft.

That was a very dramatic moment for me, and within days we in the Defense Department had set in motion the entire stealth fighter program. First, we put the program under the heaviest possible security. Second, we set in motion a large budget to ensure adequate funding all the way through. When I left office in 1981, that program was secure. We had solid funding for two squadrons of F-117s, and I knew what it could do. I had less faith that the Air Force would be able to successfully develop the tactics and armaments to maximize the stealth's tactical advantages — but the Air Force proved me wrong. They did a superb job in developing smart bombs designed just for the F-117, and they spent many months evolving the kinds of tactics that made the aircraft so extraordinarily successful in Desert Storm.

Basically, the F-117s were the enablers. They went in and destroyed or neutralized most of the Iraqi command and control centers so that nonstealth attack aircraft like the F-15s and F-16s could operate at will against targets where normally they could expect to suffer substantial losses. The F-117s made the job of our conventional aircraft infinitely safer and much easier.

So I know what the Skunk Works can do and is capable of doing, and now on my second tour at the Pentagon, I have new and greater responsibilities than the first time around and a whole new set of challenges that are, in many ways, far more complex than those we faced in the seventies. I am well aware, for example, of the overwhelming need to keep alive in an era of diminished budgets our capability for rapid development of new, experimental aircraft, which has been the forte of the Skunk Works since its inception. This is a capability I will strive to maintain in the face of broad-scale defense cutbacks. In fact, continued support of such research and development funding is uppermost among my concerns, and we have been able to maintain a reasonable level of funding despite the cutbacks and downsizing generally underway in the defense establishment.

The administration is determined to maintain the kinds of advanced prototyping and advanced testing in building experimental aircraft or weapons systems that can actually be used in small numbers in combat situations — to learn how to most effectively use them and improve them — before larger investments are made. The Skunk Works, in part, pioneered many of these techniques while building the stealth fighter.

There are three or four other small R & D operations doing advanced work other than the Skunk Works. Northrop, for instance, has an able group working on advanced aircraft and missiles; Martin Marietta has a group focusing on short-range tactical missiles. As for the Skunk Works itself, it has specialized in developing surveillance aircraft, and there is a need for a tactical reconnaissance vehicle to supply real-time eyes in combat theaters. I doubt that these will be manned airplanes. The Skunk Works has done some significant work on drones over the years, and I would see this as more economically and militarily feasible than the older U-2 and Blackbird series. I don't see those kinds of aircraft in our future: between unmanned satellites and unmanned drones, piloted reconnaissance airplanes will be squeezed out within the next five to ten years.

I don't want to compare the small R & D operations, but they all consist of similar close-knit, can-do, highly technical groups working on advanced and complex problems. They are all self-contained and do not require many people or big budgets, and as far as I am concerned, I am confident that the administration is well over the critical pass in keeping all of them afloat. I would be very concerned otherwise — if we were ignoring them or neglecting them — because I consider them to be vital to the national interest, in both the short and long term.

The Skunk Works' strength is the autonomy they have enjoyed from management and their close teamwork and partnership with their customers — both unique situations in aerospace. I have been amazed at their ability to focus technical skills in the most effective ways that really counted to problem-solving. They are the best around.

Ben R. Rich died from cancer on January 5, 1995. Ben died as he had lived — with courage, good humor, and resolve. At his request, his ashes were scattered from an airplane near his beachfront house on the California coast in Oxnard. At the moment his ashes were released, a Stealth fighter appeared out of the clouds and dipped its wings in a final salute to its creator.

INDEX

INDEX

371

146, 152–154, 157, 179, 182,
193
photos taken by, 124–125, 145,
149–150, 155, 165
amount of film, 118, 131, 167
of Cuban missile sites, 186,
187, 189–190
valued, 196
Rainbow project ("dirty bird"),
151–157, 193
refueling, 179–180
successor planned, 168, 169–
178, 192–193, 346, 350 (see
also Blackbird [SR-71 spy
plane])
test flights, 54, 55, 134–136,
137–145, 216
by foreign pilots, 141–142
secrecy of, 117–121 passim,
131, 143, 144, 145, 167
updated, 13–14, 86, 276, 335
launched from aircraft carriers, 187
postscript on, 178–191
renamed TR-1, 15, 68
in Vietnam War, 184, 185, 188
Ufimtsev, Pyotr, 19–20, 21, 29, 31
Uganda, 188
unions, 76, 216
mechanics' strike, 60
United States intelligence. See intelligence
University of Michigan, 113

Vance, Cyrus, 67, 267
Vida, Lt. Colonel Joe, 260
Vietnam War, 20, 87, 100, 105, 291,
292
Blackbird in, 238, 239, 240
combat aircraft in, 234–235,
237
defense spending after, 11, 73
Tet offensive, 245
U-2s in, 184, 185, 188
Vito, Carmen, 147–148

Wadkins, Colonel Jim, quoted on
Blackbird, 242–244
Wall Street Journal, 92
War Department, U.S., 111. See
also Defense Department,
U.S.
Warsaw Pact, 18
"weather research," U.S., 124,
145
Weinberger, Caspar, 96, 257
quoted on Skunk works, 344–
347
Welch, General Larry D., 240,
258, 317
quoted on stealth program, 41–
42, 97–98
West Germany, 144–145
intelligence, 136
Westinghouse Electric Corporation, 235
White, General Thomas, 228
White House Situation Room, 55
White Sands (N.M.) radar range,
stealth model tested at, 34–
35
Whitley, Colonel Alton, 87–88, 90–
91, 104
quoted on F-117A, 92–94
Whiz Kids, 14
World War II, 40, 109, 110, 150,
171, 212, 272
radar in, 99, 146
Skunk Works started during, 9,
10, 111, 112, 114, 337
World War III, threat of, 185. See
also nuclear weapons and
threat of war
Wright Field (Ohio), 23, 51, 175,
309, 311

X-30 project (hypersonic plane),
302
XF-104 fighter, 131. See also F-104
supersonic attack fighter
(Starfighter)
XFV-1 vertical riser, 113